Introduction to Microelectronics to Nanoelectronics

Introduction to Microelectronics to Nanoelectronics

Design and Technology

Manoj Kumar Majumder
Vijay Rao Kumbhare
Aditya Japa
Brajesh Kumar Kaushik

CRC Press
Taylor & Francis Group
Boca Raton London New York

CRC Press is an imprint of the
Taylor & Francis Group, an **informa** business

First edition published 2021
by CRC Press
6000 Broken Sound Parkway NW, Suite 300, Boca Raton, FL 33487-2742

and by CRC Press
2 Park Square, Milton Park, Abingdon, Oxon, OX14 4RN

© 2021 Taylor & Francis Group, LLC
First edition published by CRC Press 2020

CRC Press is an imprint of Taylor & Francis Group, LLC

ISBN: 9780367502379 (hbk)
ISBN: 9781003049203 (ebk)

Typeset in Times LT Std
by KnowledgeWorks Global Ltd.

Contents

List of Figures

CHAPTER 1 SEMICONDUCTOR PHYSICS AND DEVICES

CHAPTER 2 VLSI SCALING AND FABRICATION

CHAPTER 3 MOSFET MODELING

CHAPTER 4 COMBINATIONAL AND SEQUENTIAL DESIGN IN CMOS

CHAPTER 5 ANALOG CIRCUIT DESIGN

CHAPTER 6 DIGITAL DESIGN THROUGH VERILOG HDL

CHAPTER 7 VLSI INTERCONNECT AND IMPLEMENTATION

CHAPTER 8 VLSI DESIGN AND TESTABILITY

CHAPTER 10 NANOSCALE TRANSISTORS

List of Tables

CHAPTER 10 NANOSCALE TRANSISTORS

List of Abbreviations

CHAPTER 1 SEMICONDUCTOR PHYSICS AND DEVICES

BJT	Bipolar Junction Transistor
FET	Field-Effect Transistor
AC	Alternating Current
DC	Direct Current
CB	Common Base
CE	Common Emitter
CC	Common Collector
JFET	Junction Field-Effect Transistor
MOSFET	Metal Oxide Semiconductor Field-Effect Transistor
CNTFET	Carbon Nanotube Field-Effect Transistor
TFET	Tunnel Field-Effect Transistor

CHAPTER 2 VLSI SCALING AND FABRICATION

APCVD	Atmospheric Pressure Chemical Vapor Deposition
CVD	Chemical Vapor Deposition
CZ	Czochralski
EBL	Electron Beam Lithography
EGS	Electronic Grade Silicon
FET	Field-Effect Transistor
FZ	Float Zone
LM	Light Microscope
LPCVD	Low-Pressure Chemical Vapor Deposition
MBE	Molecular Beam Epitaxy
PECVD	Plasma-Enhanced Chemical Vapor Deposition
PMMA	Polymethyl Methacrylate
SEM	Scanning Electron Microscope
SOI	Silicon on Insulator
VPE	Vapor Phase Epitaxy

CHAPTER 3 MOSFET MODELING

DRAM	Dynamic Random Access Memory
GCA	Gradual Channel Approximation
ICs	Integrated Circuits
IDBL	Drain-Induced Barrier Lowering
BSIM	Berkeley Short-channel IGFET Model
IGFET	Insulated Gate Field-Effect Transistor
UC	University of California

MOS	Metal Oxide Semiconductor
MOSFET	Metal Oxide Semiconductor Field-Effect Transistor
VLSI	Very Large-Scale Integration
VTC	Voltage-Transfer Characteristics
SPICE	Simulation Program with Integrated Circuit Emphasis

CHAPTER 4 COMBINATIONAL AND SEQUENTIAL DESIGN IN CMOS

PUN	Pull-up Network
PDN	Pull-down Network
DCVSL	Differential Cascade Voltage Switch Logic
NMOS	N-channel Metal Oxide Semiconductor
PMOS	P-channel Metal Oxide Semiconductor
CMOS	Complementary Metal Oxide Semiconductor

CHAPTER 5 ANALOG CIRCUIT DESIGN

CS	Common Source
CG	Common Gate
CD	Common Drain
Op-amp	Operational Amplifiers
ADC	Analog-to-Digital Converter
DAC	Digital-to-Analog Converter
SA	Successive Approximation

CHAPTER 6 DIGITAL DESIGN THROUGH VERILOG HDL

HDL	Hardware Description Language

CHAPTER 7 VLSI INTERCONNECT AND IMPLEMENTATION

FDTD	Finite Difference Time Domain
ICs	Integrated Circuits
ITRS	International Technology Roadmap for Semiconductors
MFP	Mean Free Path
MOS	Metal Oxide Semiconductor
MOSFET	Metal Oxide Semiconductor Field-Effect Transistor
VLSI	Very Large-Scale Integration

CHAPTER 8 VLSI DESIGN AND TESTABILITY

ATPG	Automatic Test Pattern Generation
BIST	Built-in Self-Test
CUT	Circuit Under Test
DFT	Design For Testability
DUT	Device Under Test
ICs	Integrated Circuits

LFSR	Linear Feedback Shift Register
PCB	Printed Circuit Board
PO	Primary Output
ROM	Read Only Memory
SFF	Scanned Flip-Flop
SOC	System on Chip
TC	Test Control
VIA	Vertical Interconnect Access
VLSI	Very Large-Scale Integration

CHAPTER 9 NANOMATERIALS AND APPLICATIONS

AFM	Atomic Force Microscopy
CNT	Carbon Nanotube
CVD	Chemical Vapor Deposition
DIL	Driver-Interconnect-Load
DOS	Density of State
EM	Electromigration
ESC	Equivalent Single Conductor
FDTD	Finite Difference Time Domain
GNR	Graphene Nanoribbon
ITRS	International Technology Roadmap for Semiconductor
MCB	Mixed CNT Bundle
MFP	Mean Free Path
MLGNR	Multi-Layered Graphene Nanoribbon
MTL	Multi-Conductor Transmission Line
MWCNT	Multi-Walled Carbon Nanotube
SLGNR	Single-Layered Graphene Nanoribbon
SWCNT	Single-Walled Carbon Nanotube
TEM	Transmission Electron Microscopy
VLSI	Very-Large-Scale Integration

CHAPTER 10 NANOSCALE TRANSISTORS

ITRS	International Technology Roadmap for Semiconductors
DRAM	Dynamic Random Access Memory
NCFET	Negative Capacitance Field-Effect Transistor
CNFET	Carbon Nanotube Field-Effect Transistor
GNR	Graphene Nanoribbons
BTBT	Band-to-Band Tunnelling
CNT	Carbon Nanotube
SWCNT	Single-Walled Carbon Nanotube
MWCNT	Multi-Walled Carbon Nanotube
MTJ	Magnetic Tunnel Junction
GSHE	Giant Spin Hall Effect

Preface

Very large-scale integration (VLSI) is a level of integration wherein millions of transistors and discrete components with their interconnections are integrated into a semiconductor substrate. Since the invention of the integrated circuit in 1959, its manufacturing process evolved several stages from small-scale integration (SSI) to medium-scale integration (MSI), and finally came up with VLSI in the late 1980s. Advancement in the VLSI technology has led to the development of high-speed complex electronic circuits in the deep submicron and nanoscale regime. Nanotechnology or nanoscience cannot be viewed as a discipline similar to electronics, semiconductor technology, chemical technology, biotechnology, etc., which represent convergence of disciplines in some sense. It implies that the readers of this area should be in a position to pass over the boundaries of all disciplines and come up with a new area of emerging electronics.

Over the past decades, due to shrinking feature size and increasing clock frequency, the interconnections in the VLSI chip have primarily played an important role in determining the overall performance. In the recent research scenario, the interconnect delay dominates over the gate delay. With ever-increasing lengths inside a chip, global interconnects are prone to large interconnect delays, signal integrity issues, and higher current densities. Therefore, most of the conventional materials (such as Al or Cu) are susceptible to electromigration due to high current density that substantially affects the reliability of high-speed VLSI circuits. To avoid such problems, several nonsilicon or emerging devices are currently advocated as prospective material solutions in current and future nanotechnologies. Therefore, this book presents state-of-the art technology solutions for current-edge VLSI and nanoscale technology. In this book, a complete demonstration of electronics designing has been presented starting with the early-stage semiconductor devices and applications, and then moving on to the problem of scaling and VLSI fabrication, MOSFET modelling aspects, analog and digital VLSI designing, FPGA implementation using Verilog, VLSI testing and reliability, recent on-chip interconnect problems, emerging technologies, and nonsilicon (nanoscale) transistors, etc.

The detailed rundown of the subject depicted in each chapter is as follows:

CHAPTER 1 SEMICONDUCTOR DEVICES AND APPLICATIONS

This chapter starts with the semiconductor device physics and covers basic electronic devices and their evolution. Apart from this, in order to enhance the reader's understanding, emerging device technologies like fin field-effect transistors (FinFETs), Tunnel FETs (TFETs), and carbon nanotube field-effect transistors (CNTFETs) are also introduced. The functionality of electronic devices requires a basic understanding of electrons movement and physics involved in it. The movement of electrons and the probability of finding them in solids led to the identification of the difference between conductors, semiconductors, and insulators. Electrons and holes as the majority charge carriers decide whether the material is a P-type or an N-type semiconductor. A diode, as the first two-terminal device, works as a switch and shows a wide range of applications by using *PN*-junction mechanism. Later, three-terminal devices like bipolar junction transistor (BJT) and junction gate field-effect transistor (JFET) are introduced and their characteristics are analyzed. Further, metal oxide silicon field effect transistor (MOSFET) as a junctionless device is introduced to overcome the limitations of BJT and JFET. Finally, the emerging device technologies beyond MOSFET are discussed.

CHAPTER 2 VLSI SCALING AND FABRICATION

In recent years, the fabrication house has played a major role in the VLSI industry to design next-generation high-speed, high-performance integrated circuits (ICs). For a better understanding of fabrication aspects, this chapter briefly explains the history and concept of VLSI scaling from the scratch. Based on Moore's law, the journey of ICs including analog and digital circuits has been explored. To understand the fabrication process, it starts with a brief overview of the silicon wafer, and consequently demonstrates the production of raw material with detailed major fabrication steps involved in the VLSI industry. Furthermore, the basic idea of the complementary metal oxide semiconductor (CMOS) technology is also incorporated in this chapter to have the understanding of the *N*- and *P*-well CMOS process. In addition, twin-tube process of the CMOS technology is explored in detail.

CHAPTER 3 MOSFET MODELING

The primary focus of this chapter is to provide a basic understanding of MOSFET modeling. The MOSFET is the fundamental backbone of the CMOS digital integrated circuits. Compared to BJT, the fabrication process is less complex and requires less processing step and silicon area compared to BJT. In this chapter, a detailed investigation of basic structure and electrical behavior of NMOS (*n*-channel MOS) and PMOS (*p*-channel MOS) are examined to understand the advantages of the MOS transistor that is widely used as a switching device in LSI and VLSI circuits. Subsequently, the MOSFET model is required to understand the circuit simulation that is classified as the (i) DC model or a steady state model, wherein the applied voltage remains constant and does not vary with time; (ii) dynamic or AC model, wherein the applied voltage does not remain constant but varies with time. In this chapter, the different regions of device operation of a DC MOS transistor and dynamic models are discussed. In addition , this chapter also includes the implementation of the MOSFET model in Berkeley SPICE2G and a higher version. The four different MOSFET models of Berkeley SPICE are investigated for varying complexity and accuracy.

CHAPTER 4 COMBINATIONAL AND SEQUENTIAL DESIGN IN CMOS

This chapter primarily introduces the CMOS inverter design and analyses its performance. Static and dynamic behavior of the CMOS inverter is studied. In static behavior, the voltage transfer characteristic of CMOS is presented and is used to define the noise margins. Dynamic behavior of the inverter is elaborated with propagation delay and power consumption metrics. By using both power and delay, the energy metric of CMOS inverter is introduced. Later, CMOS-based combinational and sequential circuits are introduced based on which the static CMOS gates are designed. Moreover, the ratioed, pass transistor, and dynamic logic are introduced that are alternative to static CMOS designs. Metrics for a sequential design like setup time, hold time, and bistability principle are introduced. Later, CMOS-based SR latch and flip-flop circuit designs are presented. These designs are further utilized to develop D flip-flop.

CHAPTER 5 ANALOG CIRCUIT DESIGN

This chapter starts with the motivation behind the learning of CMOS analog design. Later, MOSFET physics is described and device analog metrics are introduced. MOSFET-based single- and multi-stage amplifiers are introduced and different analog metrics like gain, input, and output resistances are analyzed. Different current mirror circuits are considered and analyzed. Difference between single-ended and differential circuits is studied. This concept further introduces basic differential pair and analyzes its gain. Later, CMOS based op-amp and different stages of it are introduced. The frequency response of op-amp is demonstrated to get basic analog metrics. CMOS-based comparators are designed and analyzed. Finally, different CMOS-based analog-to-digital converters and digital-to-analog converters are presented.

CHAPTER 6 DIGITAL DESIGN THROUGH VERILOG HDL

This chapter focuses on Verilog hardware description language (HDL) and explains the features of the hardware description language. First, basic concepts including lexical tokens, data types, and operators are introduced. Then, the basic syntax of the module and the test bench is presented with examples. Different modeling styles of Verilog including structural, dataflow, and behavioral styles are reviewed with examples. Initial and always blocks are presented and the difference between them has been analyzed. Combinational and sequential circuits are exemplified to explain the different modeling styles. Finally, exercises are provided to enhance the reader's understanding and problem-solving abilities.

CHAPTER 7 VLSI INTERCONNECT AND IMPLEMENTATION

This chapter critically presents in-depth analysis to understand the behavior of interconnects in the modern VLSI IC technology. It starts with the conventional interconnect technique and problem arises with the conventional material. Based on the technology scaling, different types of electrical circuit modeling has been discussed in detail for resistive and capacitive parasitics. These interconnect parasitics primarily affects the signal integrity, power dissipation, and crosstalk performance of an IC. Furthermore, a number of delay models present a comprehensive demonstration of interconnect performance and propose the next generation interconnect technology.

CHAPTER 8 VLSI DESIGN AND TESTABILITY

This chapter introduces the behavior of circuits that provides a proper diagnostic by generating a test pattern and comparing it with a golden machine. As the testing of circuit and devices in the VLSI industry is an essential step due to miniaturization of devices that increase in complexity, predefined test models are used to observe the faults and rectify them based on the appropriate models. In the current VLSI industry, the designers add some extra circuitry to lower the cost of tests, which increases the testing procedure, enhances the diagnostic process, and makes them more efficient and accurate.

CHAPTER 9 NANOMATERIALS AND APPLICATIONS

This chapter is designed to make the readers aware about the technology advancement to help them work in the area of nanomaterials. This chapter begins with the basic knowledge of carbon nanotubes (CNTs) and graphene nanoribbons (GNRs) that are carbon allotropes. These materials have hexagonal arrangements of carbon atoms and possess unique electrical, mechanical, thermal, and optical properties. For a better understanding, the use of CNTs and GNRs is demonstrated in different fabrication approaches to produce high-quality nanomaterials. Afterward, equivalent electrical modeling of CNT and GNR is presented based on the geometrical structure. The performance in terms of propagation delay, crosstalk, and power dissipation is analyzed using industry standard HSPICE to understand its importance as the next-generation interconnect application. In addition, active and passive shielding for CNT and GNR are demonstrated to observe the impact of shielding to enhance the performance in deep nanotechnology. Subsequently, other applications of nanomaterials in nanosensors, nanofilter, bullet-proof combat jacket, fire extinguisher, medical technology, etc are discussed.

CHAPTER 10 NANOSCALE TRANSISTORS

This chapter presents different applications by exploring the characteristics of emerging device technologies that are alternative to the existing CMOS VLSI technology. Recently, many beyond-CMOS devices such as Tunnel FETs (TFETs), Negative Capacitance FETs (NCFETs), Carbon Nanotube FETs, Graphene FETs, and STT-MRAMs have emerged with promising speed and ultra-low power consumption. The structure and different fabrication aspects of these nonsilicon transistors are explored in this chapter. Additionally, a comparative study is also presented to ensure several benefits of the nonsilicon transistors with the existing CMOS technology.

Authors

Manoj Kumar Majumder received his PhD from Microelectronics and VLSI group at Indian Institute of Technology, Roorkee, India. Currently, he works as an assistant professor in the Department of Electronics and Communication Engineering, IIIT Naya Raipur, Chhattisgarh. He has authored more than 30 papers in peer-reviewed international journals and more than 45 papers in international conferences. He has co-authored a book titled *Carbon Nanotobe-Based VLSI Interconnects-Analysis and Design* (New York, NY, USA: Springer, 2014) and also a book chapter published by CRC Press. His current research interests include the area of graphene-based low-power VLSI devices and circuits, on-chip VLSI interconnects and through-silicon vias. Dr. Majumder is associated with different academic and administrative activities of different positions in IIIT, Naya Raipur. He is an active senior member of IEEE and IEEE Electron Device Society. He has delivered technical talks at different conferences in India and abroad. He is also an active reviewer of *IEEE Transactions on Electromagnetic Compatibility, IEEE Transactions on Nanotechnology, IEEE Electron Device Letters, Microelectronics Journal, IET Micro & Nano Letters*, and other Springer journals. He is a member of many expert committees constituted by government and non-government organizations. His name has been listed in Marquis Who's Who in the World®.

Vijay Rao Kumbhare received his B.Tech degree in Electronics and Tele-communication Engineering from the National Institute of Technology, Raipur, India in 2008, and M.Tech. degree with specialization in Telecommunication System Engineering (TSE) under Electronics and Electrical Communication Engineering Department from the Indian Institute of Technology, Kharagpur, West Bengal, India, in 2011. He worked as an Assistant Professor for more than 5 years. He is currently working toward his Ph.D. degree from Dr. Shyama Prasad Mukherjee International Institute of Information Technology Naya Raipur, India. He has published 1 IEEE transaction, 5 journals and 7 reputed conference papers. He is an active IEEE student member. He has attended several workshops and conferences, and is an active reviewer of reputed IET Biotechnology. His current research interests are in the area of graphene nanoribbon, carbon nanotube, and optical-based on-chip VLSI interconnects, and emerging nanomaterials.

Aditya Japa received his B.Tech. degree in Electronics and Communication Engineering from Sree Chaitanya College of Engineering, Karimnagar (J.N.T.U Hyderabad), Telangana, India, in 2012 and M.Tech. degree in VLSI Design from Vignan's University, Andhra Pradesh, India, in 2015. He was a JRF under a DST project titled "Design, Analysis and Benchmarking of Energy-Efficient Hetero-Junction Tunnel FET-based Digital, Analog and RF Building Blocks" during 2015–2016. He is currently pursuing his Ph.D. in Electronics and Communication Engineering from DSPM International Institute of Information Technology, Naya Raipur, India.

His current research interests include hardware security subsystems like true random number generators (TRNG) and physical unclonable functions (PUF), emerging transistor technologies (Tunnel FETs), ultra-low power/energy efficient sensor readout circuits, and VLSI design etc.

Brajesh Kumar Kaushik received his Doctorate of Philosophy (Ph.D.) in 2007 from the Indian Institute of Technology, Roorkee, India. He joined the Department of Electronics and Communication Engineering, Indian Institute of technology, Roorkee, as an assistant professor in December 2009; and since April 2014 he has been an associate professor. He has extensively published in several national and international journals and conferences. He is a reviewer of many international journals belonging to various publishers including IEEE, IET, Elsevier, Springer, Taylor & Francis, Emerald, ETRI, and PIER. He has served as General Chair, Technical Chair, and Keynote Speaker of many reputed international and national conferences. Dr. Kaushik is a senior member of IEEE and a member of many expert committees constituted by Government and non-Government organizations. He holds the position of Editor and Editor-in-Chief of various journals in the field of VLSI and microelectronics, such as *International Journal of VLSI Design & Communication Systems* (*VLSICS*), AIRCC Publishing Corporation. He also holds the position of Editor of *Microelectronics Journal* (*MEJ*), Elsevier Inc.; *Journal of Engineering, Design and Technology* (*JEDT*), Emerald Group Publishing Limited; *Journal of Electrical and Electronics Engineering Research* (*JEEER*) and Academic Journals. He has received many awards and recognitions from the International Biographical Center (IBC), Cambridge. His name has been listed in Marquis Who's Who in Science and Engineering® and Marquis Who's Who in the World®. Dr. Kaushik has been conferred with Distinguished Lecturer award of IEEE Electron Devices Society (EDS) to offer EDS Chapters with a list of quality lectures in his research domain.

Acknowledgments

First, we acknowledge the valuable and highly regarded contributions from Prof. Ashok K. Srivastava, Louisiana State University, USA, and Prof. Pradeep K. Sinha, Director, IIIT Naya Raipur, INDIA, for their original and innovative thought behind the writing of this book. They have always been cordial, thoughtful, trustworthy, and encouraging in every nook and cranny throughout the preparation of the book. With their astute and insightful intellect, maestro aptitude, and deep knowledge, they have duly steered the effort in a steady direction. We express our sincere thanks to Prof. Sanjeev Manhas, IIT Roorkee; Prof. Sudeb Dasgupta, IIT Roorkee; and Dr. Ramesh Vaddi, SRM University, A.P., for enlightening us with their valuable suggestions and broadening our perspective. We are also thankful to our students Majeti L. V. S. S. Krishna Koushik, and Velpula Sanjeet for constructive technical discussions at various stages of chapter preparation that made it possible to complete this book on time.

The authors recognize the invaluable inputs made by the ever-active team of commissioning editors, content developers, and editors. We are very much thankful to the anonymous editors and reviewers of the book for in-depth and critical analysis of the content that helped us to substantially improve the quality of the book. We are really grateful to Prof. Hong Xiao, Prof. Hongjie Die, Prof. Kirsten Moselund, Prof. Joerg Appenzeller, and Prof. Jeong-Sun for their acknowledgment and for providing the permission toward the reproduction of various fabricated figures from their research work. We also acknowledge the contribution of various publishers such as CRC Press, IEEE-Willey, and Pearson for their copyright permission in various cases.

Last but certainly not the least, we are highly indebted to our families whose affection, love, and blessings have been a source of light for the successful completion of our book. Without their encouragement and good wishes, we could not have come to this stage. We offer our sincere gratitude to the Almighty for giving us the right inspiration at the right time, blessing us with all the good fortune and company of the right people who helped us in moving toward our aspiration in life.

This book is dedicated to our parents and loving family.

Manoj Kumar Majumder
Vijay Rao Kumbhare
Aditya Japa
Brajesh Kumar Kaushik

1 Semiconductor Physics and Devices

1.1 INTRODUCTION

Semiconductors have revolutionized the field of electronics and play a prominent role in our day-to-day life. The seed of development of these modern solid-state semiconductors dates back to early 1930s. Each electronic device that we see around is made up of semiconductors. We wouldn't have been able to achieve these remarkable results without them.

1.1.1 CONDUCTION IN SOLIDS

The form in which matter exists is called the state of matter. Matter exists in many distinct states of which three states are well known and important. They are solids, liquids, and gases, and other states include plasma, Bose–Einstein condensates, degenerate matter, photonic matter, etc. These other states occur only in extreme conditions of pressure, temperature, and energy. The main difference in the structure of each state lies in the densities of the particles. The density of particles is highest in solids and lowest in gases. Figure 1.1 shows the alignment of particles in different states of matter and the processes through which we can convert one state of matter to another.

Solids are further classified into three types based upon the distance between their valence band and conduction band. This distance is called the bandgap [1]. The electrical conductivity of the substance depends upon the bandgap of the material. We say that the substance is able to conduct if there are free electrons in the conduction band. When energy is supplied to the elements, the electrons in the valence band get excited and jump up into the conduction band, allowing the passage of electricity through the substance. In conductive materials, no bandgap exists, due to which electrons can move easily between their valence band and the conduction band. Unlike conductors, insulators have a huge bandgap between the conduction and the valence band. The valence band remains full since no movement of electrons occurs, and as a result, the conduction band remains empty as well. In semiconductor materials, the bandgap between the conduction band and the valence band is smaller. At room temperature, there is enough energy accessible to displace a few electrons from the valence band into the conduction band. As temperature increases, the conductivity of a semiconductor material increases. Figure 1.2 shows the bandgaps in conductors, semiconductors, and insulators, respectively.

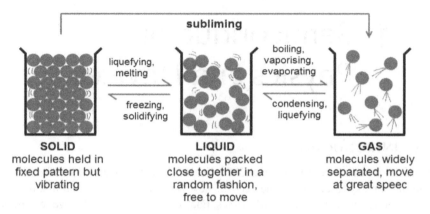

FIGURE 1.1   Alignment of particles in different states of matter.

1.1.2   CONDUCTORS, INSULATORS, AND SEMICONDUCTORS

This subsection provides a detailed description of different material properties.

1.1.2.1   Conductors

The substances that allow electricity to pass through them are called conductors. Metals such as gold, silver, and copper are good examples of conductors. These substances have free electrons in their outermost orbit. There is no or very little distance between the conduction band and the valence band.

Properties of Conductors

- Conductors have high electrical and thermal conductivities.
- In steady states, they obey Ohm's law.
- They have a positive temperature coefficient, i.e., their resistance increases with an increase in temperature.
- They obey Wiedemann–Franz law, according to which the ratio of thermal and electrical conductivities at a given temperature is the same for all metals and is proportional to the absolute temperature T, as shown below:

$$\frac{K}{\sigma} \propto T \tag{1.1}$$

where K is the thermal conductivity of the metal, σ is the electrical conductivity of the metal, and T is the temperature.

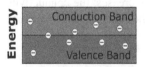 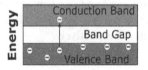 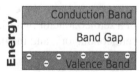

FIGURE 1.2   Band diagrams of conductors, semiconductors, and insulators.

1.1.2.2 Insulators

The substances that do not allow electricity to pass through them are called insulators. They are also known as non-conducting substances. Glass, mica, and quartz are good examples of insulators. The outermost orbits of these substances are saturated, and hence no free electrons are present. The distance is extremely long between the valence band and the conduction band of these substances.

Properties of Insulators

- Resistance is the ability to impede the electric current. Insulators have extremely high resistance. The resistance is in the order of ohms in metals, whereas in insulators, it varies from several kilo-ohms to mega ohms.
- The conductivity of the insulators does not change with temperature.
- Insulators, after a specific voltage, lose their insulating properties and start conducting. This voltage is known as the breakdown voltage. This is also called the dielectric strength of the material. Insulators have very high breakdown voltages.
- Insulators have very high permeability.

1.1.2.3 Semiconductors

These substances have their conductivity between the conductors and the insulators. Their resistivity is higher than that of a conductor but lesser as compared to an insulator. The distance between the valence band and the conduction band is intermediate to that of conductors and insulators.

Properties of Semiconductors

- Semiconductors have negative temperature coefficient of resistance. Therefore, the electrical resistance of a semiconductor reduces with an increase in the temperature.

$$R = A \times e^{\frac{-B}{T}} \tag{1.2}$$

  where R represents the resistance of the semiconductor, T is the temperature, and A and B are the constants.
- The electrical conductivity of a semiconductor can be increased enormously by adding a small amount of impurity. Germanium (Ge) and silicon (Si) are the most commonly used semiconductors. Germanium is used extensively in early solid-state devices such as transistors, but it is now being replaced by silicon because of its abundance. The properties of conductors, semiconductors, and insulators are summarized in **Table 1.1**.

1.1.3 P-TYPE AND N-TYPE SEMICONDUCTORS

Before moving further, we should be able to understand the basics of semiconductors. As explained earlier, the conductivity of the semiconductors lies between that of the conductors and the insulators. On their own, silicon and germanium are classed

TABLE 1.1

A Comparison of Conductor, Semiconductor, and Insulator Material properties

Characteristics	Conductor	Semiconductor	Insulator
Conductivity	High	Moderate	Low
Resistivity	Low	Moderate	Very high
Forbidden gap	No forbidden gap	Small forbidden gap	Large forbidden gap
Temperature coefficient	Positive	Negative	Negative
Conduction	Large number of electrons for conduction	Very small number of electrons for conduction	Moderate number of electrons for conduction
Conductivity value	Very high 10^{-7} mho/m	Between those of conductors and insulators, i.e., 10^{-7} mho/m to 10^{-13} mho/m	Negligible like 10^{-13} mho/m
Resistivity value	Negligible; less than 10^{-5} Ω-m	Between those of conductors and insulators, i.e., 10^{5} Ω-m to 10^{5} Ω-m	Very high; more than 10^{5} Ω-m
Current flow	Due to free electrons	Due to holes and free electrons	Due to negligible free electrons
Number of carriers at normal temperature	Very high	Low	Negligible
Band overlap	Both conduction and valence bands are overlapped	Both bands are separated by an energy gap of 1.1 eV	Both bands are separated by an energy gap of 6–10 eV
Zero Kelvin behavior	Acts like a superconductor	Acts like an insulator	Acts like an insulator
Formation	Formed by metallic bonding	Formed by covalent bonding	Formed by ionic bonding
Valence electrons	One valence electron in outermost shell	Four valence electrons in outermost shell	Eight valence electrons in outermost shell
Examples	Copper, mercury, aluminum, silver	Germanium, silicon	Wood, rubber, mica, paper

as intrinsic semiconductors that exhibit their purity. However, by controlling the amount of impurities added to this intrinsic semiconductor material, it is possible to control its conductivity. Various impurities called donors or acceptors can be added to this intrinsic material to produce free electrons or holes, respectively. This process of adding donor or acceptor atoms to semiconductor atoms is known as doping. The doped silicon is no longer pure, whereas these donor and acceptor atoms are collectively referred to as "impurities," and by doping these silicon materials with a sufficient number of impurities, we can turn them into an *N*-type or *P*-type semiconductor materials. The most commonly used semiconductor material is silicon. Figure 1.3 shows the structure and lattice of a "normal" pure crystal of silicon.

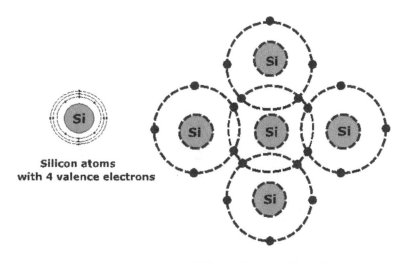

**Silicon atoms
with 4 valence electrons**

Silicon Crystal Lattice

FIGURE 1.3   Structure and lattice of a "normal" pure crystal of silicon.

Silicon has four valence electrons in its outermost shell that shares its neighboring silicon atoms to form full orbitals of eight electrons. The structure of the bond between the two silicon atoms is such that each atom shares one electron with its neighbor making the bond highly stable.

As there are few free electrons available to move around the silicon crystal, crystals of pure silicon (or germanium) are therefore good insulators, i.e., they possesses extremely low or high value of resistances. Silicon atoms are arranged in a definite symmetrical pattern making them a crystalline solid structure. A crystal of pure silica (silicon dioxide or glass) is generally said to be an intrinsic crystal (it has no impurities) and therefore has no free electrons. But simply connecting a silicon crystal to a battery supply is not enough to extract an electric current from it. In order to do that, we need to create "positive" and "negative" poles within the silicon crystal allowing the electrons and hence the electric current to flow out of silicon. These poles are created by doping the silicon with certain impurities.

1.1.3.1   *N*-Type Semiconductors

In order to improve the conductivity, we need to introduce an impurity atom such as arsenic, antimony, or phosphorus into the crystalline structure making it extrinsic (impurities are added). These atoms have five outer electrons in their outermost orbital to share with neighboring atoms and are commonly called "pentavalent" impurities. This allows four out of the five orbital electrons to bond with its neighboring silicon atoms leaving one "free electron" to become mobile when an electrical voltage is applied (electron flow). As each impurity atom "donates" one electron, pentavalent atoms are generally known as "donors."

Antimony (symbol Sb) and phosphorus (symbol P) are frequently used as pentavalent additives to silicon. Figure 1.4 shows the structure and lattice of the donor atom impurity. Antimony has 51 electrons arranged in five shells around its nucleus

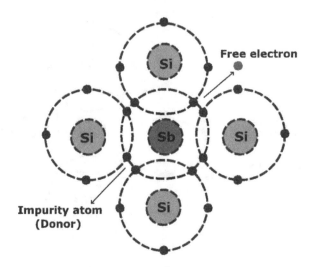

n-type semiconductor

FIGURE 1.4   Structure and lattice of the donor atom impurity.

with the outermost orbital having five electrons. The resulting semiconductor basic material has an excess of current-carrying electrons, each with a negative charge, and is therefore referred to as an *N*-type material with the electrons called "Majority Carriers," while the resulting holes are called "Minority Carriers." When stimulated by an external power source, the electrons freed from the silicon atoms by this stimulation are quickly replaced by the free electrons available from the doped Antimony atoms. However, this action still leaves an extra electron (the freed electron) floating around the doped crystal, making it negatively charged. Therefore, a semiconductor material is classed as *N*-type when its donor density is greater than its acceptor density; in other words, it has more electrons than holes, thereby creating a negative pole, as shown in Figure 1.4.

1.1.3.2   *P*-Type Semiconductors

If we go the other way, and introduce a "Trivalent" (3-electron) impurity into the crystalline structure, such as aluminum, boron, or indium that have only three valence electrons available in their outermost orbital, the fourth closed bond cannot be formed. Therefore, a complete connection is not possible, giving the semiconductor material an abundance of positively charged carriers known as holes in the structure of the crystal wherein the electrons are effectively missing. As a result, a hole exists in the silicon crystal, and a neighboring electron is attracted to it and will try to move into the hole to fill it. However, the electron filling the hole leaves another hole behind as it moves. This in turn attracts another electron that in turn creates another hole behind it, and so forth by giving the appearance that the holes are moving as a positive charge through the crystal structure (conventional current flow). This movement of holes results in a shortage of electrons in the silicon turning the entire doped crystal into a positive pole. As each impurity atom generates a

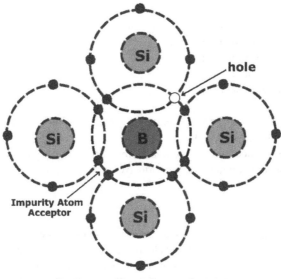

P-type Semiconductor

FIGURE 1.5 Structure and lattice of the acceptor impurity atom boron.

hole, trivalent impurities are generally known as "**Acceptors**" as they are continually "accepting" extra or free electrons. **Boron** (symbol B) is commonly used as a trivalent additive as it has only five electrons arranged in three shells around its nucleus with the outermost orbital having only three electrons. Figure 1.5 presents the structure and lattice of the acceptor impurity atom boron. The doping of boron atoms causes conduction that consist mainly of positive charge carriers resulting in a **P-type** material with the positive holes called "Majority Carriers," while the free electrons are called "Minority Carriers." Therefore, a semiconductor basics material is classified as P-type when its acceptor density is greater than its donor density. Therefore, a P-type semiconductor has more holes than electrons.

1.1.4 SEMICONDUCTOR CONDUCTIVITY

In an intrinsic semiconductor, the concentration of electrons and holes is identical. The electrons and holes move in opposite directions under an electric field "E." The total current density J within the intrinsic semiconductor is given by

$$J = J_n + J_p$$

$$= q.n.\,\mu_n.E + p.n.\,\mu_p.E$$

$$= \left(n.\mu_n + p.\mu_p\right)qE$$

$$J = \sigma E \qquad (1.3)$$

where J_P = current density due to free holes

J_n = current density due to free electrons

n = concentration of electrons

p = concentration of holes

E = electric field

q = charge of electrons

Here, σ is the conductivity of semiconductor, which is equal to $\left(n.\mu_n + p.\mu_p\right)q$.

From the above equation (1.3),

For N-type of semiconductor, the conductivity, $\sigma = n \cdot q \cdot \mu_n \ (p \ll n)$ $\qquad$ (1.4)

For P-type of semiconductor, the conductivity, $\sigma = p \cdot q \cdot \mu_p \ (n \ll p)$ $\qquad$ (1.5)

1.2 DIODE

Diodes are the simplest, yet most used electronic device in the semiconductor world. They act as building blocks for several electronic devices. Diode was originally called a rectifier because of its ability to convert alternating current (AC) to direct current (DC). It was renamed as diode in 1919 by English physicist William Henry Eccles who coined the term from the Greek root *di*, meaning "two," and *ode*, a shortened form of "electrode." Earlier diodes were made from selenium (Se) and germanium (Ge). However, they have been completely replaced by silicon (Si) in the recent years.

1.2.1 DIODE STRUCTURE AND CHARACTERISTICS

A diode is a semiconductor device that allows the flow of current only in one direction. It consists of two electrodes called cathode (negative terminal) and anode (positive terminal), as shown in Figure 1.6. When voltage polarity on the anode side is positive as compared to that on the cathode side, the diode conducts and is considered a low-value resistor. If voltage polarity at the anode side is more negative as compared to that on the cathode side, the diode is said to be in reverse-biased mode and it does not conduct. There are many types of diodes of which *PN* diode and Zener diode are the most important due to their applications. Here, we will explicitly explain about the *PN* diode.

1.2.2 *PN* DIODE STRUCTURE

The newly doped N-type and P-type semiconductor materials do very little on their own, as they are electrically neutral. However, if we join (or fuse) these two semiconductor materials together, they behave in a very different way merging together and producing what is generally known as a "*PN* Junction [2]." When the N-type

FIGURE 1.6   Diode symbol.

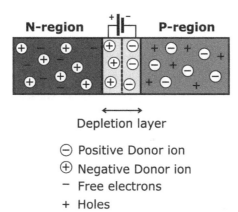

Depletion layer

⊖ Positive Donor ion
⊕ Negative Donor ion
− Free electrons
+ Holes

FIGURE 1.7 *PN* Diode internal structure at zero bias.

semiconductor and *P*-type semiconductor materials are first joined together, a large density gradient exists between both sides of the *PN* junction. As a result, some of the free electrons from the donor impurity atoms begin to migrate across this newly formed junction to fill up the holes in the *P*-type material producing negative ions. However, because the electrons have moved across the *PN* junction from the *N*-type silicon to the *P*-type, they leave behind positively charged donor ions (N_D) on the negative side. Now the holes from the acceptor impurity migrate across the junction in the opposite direction into the region where there are large numbers of free electrons.

As a result, the charge density of the *P*-type along the junction is filled with negatively charged acceptor ions, and the charge density of the *N*-type along the junction becomes positive, as shown in Figure 1.7. This charge transfer of electrons and holes across the *PN* junction is known as diffusion. The width of these *P* and *N* layers depends on how heavily each side is doped with acceptor density N_A and donor density N_D, respectively. This process continues back and forth until the number of electrons, which have crossed the junction, have a large enough electrical charge to repel or prevent any more charge carriers from crossing over the junction. Eventually, a state of equilibrium (electrically neutral situation) will occur producing a "potential barrier" zone around the area of the junction as the donor atoms repel the holes and the acceptor atoms repel the electrons. Since no free charge carriers can rest in a position where there is a potential barrier, the regions on either side of the junction now become completely depleted of any more free carriers in comparison to the *N*- and *P*-type materials further away from the junction. This area around the *PN* junction is now called the depletion layer. The total charge on each side of a *PN junction* must be equal and opposite to maintain a neutral charge condition around the junction. If the depletion layer region penetrates into the silicon by a distance of D_P for the positive side, and a distance of D_N for the negative side giving a relationship between the two as equation (1.6):

$$D_P N_A = D_N N_D \qquad (1.6)$$

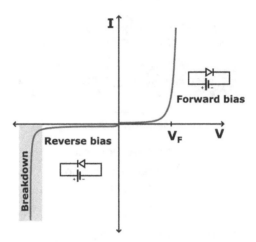

FIGURE 1.8   *PN* diode forward and reverse bias characteristics.

1.2.2.1 Forward and Reverse Bias Regions

When the diode *P*-type material is connected to a positive voltage and *N*-type material is connected with a negative voltage, the diode operates in forward bias, and the diode *i*–*v* relationship is approximated by equation (1.7).

$$i = I_S \left(e^{v/nV_t} - 1 \right) \tag{1.7}$$

When the positive voltage is applied to *N*-type material and a negative voltage is applied to *P*-type material, the diode operates in reverse bias, and in this region, zero current flows through the diode. The characteristic of diode is shown in Figure 1.8. In forward bias, the diode voltage should be more than the threshold voltage to allow current; this voltage is called cut-in voltage (V_F) of the diode. From these characteristics, it can be observed that the diode exhibits approximately zero current in reverse direction.

1.2.3 ZENER DIODE STRUCTURE

This is a *PN* junction device, wherein Zener breakdown mechanism dominates. Zener diode is always used in reverse bias and its symbol is shown in Figure 1.9. A Zener diode has the following features:

1. Doping concentration is heavy on *P* and *N* regions of the diode, compared to a normal *PN* junction diode.

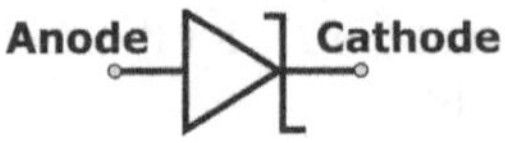

FIGURE 1.9   Zener diode symbol.

2. Due to heavy doping, depletion region width is narrow.
3. Due to narrow depletion region width, electric field intensity is high that results in Zener Breakdown mechanism.

The Zener diode operates in the reverse breakdown region and exhibits almost a constant voltage drop that can be used in voltage regulator circuits. These types of diodes are designed to be operated in the breakdown region called breakdown diodes or Zener diodes. Figure 1.10 presents the Zener diode characteristics showing constant Zener voltage in the reverse breakdown region, whereas in forward region, it shows normal *PN* diode characteristics only. Zener diode can be used in different applications like voltage regulator circuits, clipping and clamping circuits, and wave-shaping circuits.

1.2.4 DIODE APPLICATIONS

Apart from the switching characteristics, a diode can be explored in many important applications like rectifiers, clippers, and clampers.

1.2.4.1 Rectifiers

Most of the electronic circuits require a DC source of power. The circuit that converts AC to pulsating DC is called a rectifier. Generally, AC input is a sinusoidal signal, as shown in Figure 1.11, and the rectifier circuit converts this sinusoidal

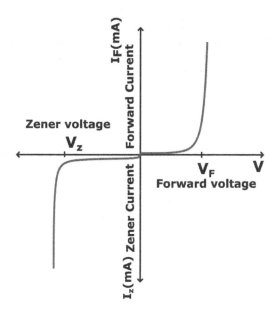

FIGURE 1.10   Zener diode characteristics.

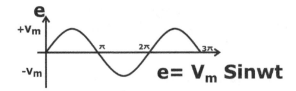

FIGURE 1.11   Sinusoidal AC input signal.

signal into a pulsating DC signal. Different types of rectifier are available, and they include:

- Half-wave rectifier
- Full-wave rectifier
    i.   Center-tapped rectifier
    ii. Bridge rectifier

1.2.4.1.1   Half-Wave Rectifier

A half-wave rectifier is designed using a single diode and a resistor connected in series and is shown in Figure 1.12. The output is measured across the resistor, during the positive half cycle of input, diode switches ON. Due to this, input passes to output. With the negative half cycle, input diode is switched OFF creating an open circuit and resulting in zero output. The output voltage here is not exactly DC and it varies between maximum sinusoidal amplitude and zero, as shown in Figure 1.12.

1.2.4.1.2   Full-Wave Center-Tapped Rectifier

This circuit consists of two diodes and a load resistor R_O. A center-tapped transformer is used to get full rectified signal at the output [3]. During the positive half cycle of input, diode D_1 switches ON and D_2 switches OFF, as shown in Figure 1.13(a). During the negative half cycle, D_2 switches ON and D_1 switches OFF, and the direction of current through the load is not changed, as shown in Figure 1.13(b). As a result,

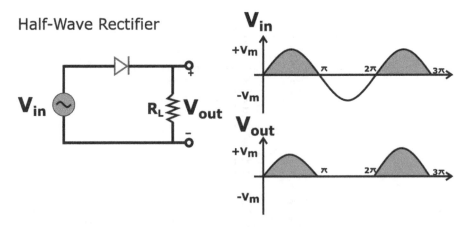

FIGURE 1.12   Half-wave rectifier circuit design with output response.

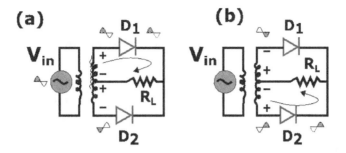

FIGURE 1.13 Full-wave rectifier circuit design (a) during positive half cycle (b) during negative half cycle.

output produces another positive half cycle, as shown in Figure 1.14. In the full-wave rectifier, identifying the center tap of transformer is a difficult task and hence a bridge rectifier is introduced.

1.2.4.1.3 Bridge Rectifier

The bride rectifier consists of four diodes connected in a bridge fashion, as shown in Figure 1.15. During the positive half cycle, D_1 and D_3 are forward biased at the same time, whereas D_2 and D_4 are connected in reverse biased. In this condition, we can observe the current direction through the load, as shown in Figure 1.15. During the negative half cycle, D_2 and D_4 are forward biased, D_3 and D_1 are reverse biased. It can be observed that in this condition, the current through this load resistance flows in the same direction. As a result, output produces positive half cycle.

1.2.4.1.4 Comparison of Rectifier Circuits

The performance of rectifier is evaluated by using certain metrics such as efficiency, ripple factor, and number of diodes used. These parameters for three different rectifier circuits are summarized in **Table 1.2**. It can be observed that bridge rectifier with its simple circuit shows high accuracy and low ripple factor value.

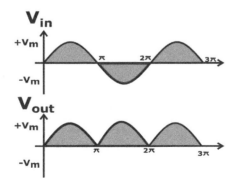

FIGURE 1.14 Full-wave rectifier output.

TABLE 1.2

Performance Comparison of Various Diode-Based Rectifiers

		Full-Wave Rectifier	
Performance parameters	Half-Wave Rectifier	Center Tap	Bridge
Efficiency	40.6%	81.2%	81.2%
Ripple factor	1.21	0.482	0.482
No. of diodes	1	2	4
Center tap necessity	No	Yes	No

1.2.4.2 Diode Logic Gates

Diode is used to construct digital logic gates that are shown in Figure 1.16. Here, both AND and OR gates are constructed using diodes.

In the AND gate, when both the inputs are high, the diodes will be in reverse bias and output will be connected to maximum voltage (V). In remaining cases of inputs, one of the diodes will be forward biased that makes the output logic zero. The Boolean notation for the circuit can be given by

$$Y = A.B$$

(a)   **(b)**

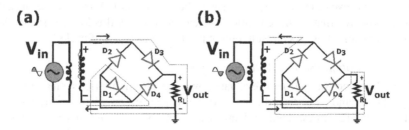

FIGURE 1.15 Bridge rectifier circuit behavior (a) during positive half cycle (b) during negative half cycle.

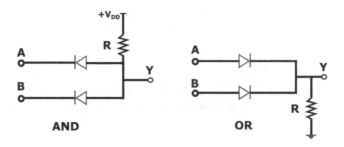

FIGURE 1.16 AND and OR logic gates using diodes.

In the case of OR gate, if one of the inputs is logic high, output shows logic high. In the remaining cases, output will be zero. Therefore, the Boolean notation for the circuit is

$$\mathbf{Y = A + B}$$

1.2.4.3   Clipping and Clamping Circuits

The diode can act as an open or closed switch depending on the biasing voltage applied. Due to this, it has been explored for different applications. The clipping and clamping function also explores the diode characteristics.

1.2.4.3.1   Clipper Circuits

A clipper circuit clips off or cuts the portion of the signal and produces the remaining signal as output. There are two types of clipper circuits, the series and parallel diode clipping circuits.

1.2.4.3.2   Series Diode Clipping Circuit

In these types of clipping circuits, the diode is connected between the input and the output terminals, as shown in Figure 1.17.

1.2.4.3.3   Operation of Clipping Circuit

A series clipper is considered to explain the operation of a circuit, as shown in Figure 1.17. Here, a sine wave is applied as input to the clipper circuits. In the positive half cycle of the input, diode is switched ON, the input passes to the output and can be measured across the output. In the negative half cycle of the input, diode is switched OFF, and the circuit becomes an open circuit. As a result, the output will be zero. Thus, this clipper circuit clips the negative half cycle of signal, as shown in Figure 1.17.

1.2.4.3.4   Parallel Diode Clipping Circuit

In this type of clippers, the diode is connected between the output terminals, as shown in Figure 1.18. The on/off state of the diode directly affects the output voltage. These types of clippers may also have a non-zero threshold voltage by addition of a voltage series with diode. The following figure illustrates the clipping process.

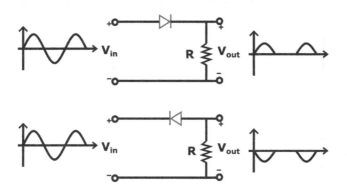

FIGURE 1.17   Series clippers using diodes.

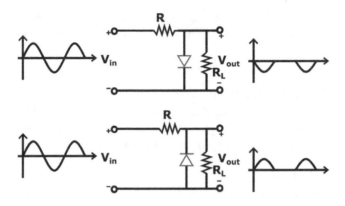

FIGURE 1.18   Parallel clippers using diodes.

1.2.4.3.5   Clamper Circuits

Clamper circuits are explored to change the DC level of a signal. In contrast to clippers circuits, clamper uses a capacitor and a diode connection. Depending on the clamping position, clampers can be classified into two types:

- Positive clamper
- Negative clamper

The behavior of the positive and negative clampers is shown in Figure 1.19; from this, it can be concluded that positive and negative clampers add the positive and negative voltages to the actual signals.

1.2.4.3.6   Operation

Generally, a clamper is an *RC* circuit that consists of a diode and a capacitor. The diode changes its configuration, i.e., open or closed switch depending on the input applied. In order to clearly understand the clamper operation, let us consider the clamper shown in Figure 1.20. This clamper consists of a capacitor in series with diode. The input to the clamper is applied to the capacitor and the output is measured

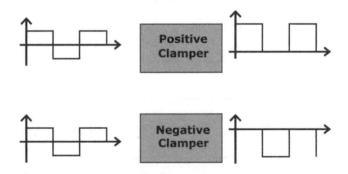

FIGURE 1.19   Block diagram of clamper design.

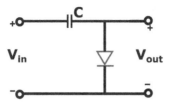

FIGURE 1.20 Clamper circuit.

across the diode, as shown in Figure 1.20. Here, we are applying a square wave as an input to the clamper. The operation can be explained in terms of the applied positive and negative half cycles at the input.

During positive half cycle of input: With positive half cycle of input, the capacitor charges to input voltage and the diode becomes forward biased. The forwarded diode acts as short circuit and the voltage across the diode will be approximately zero, as shown in Figure 1.21.

During negative half cycle of input: With negative half cycle of input, the diode becomes revere biased and acts as open circuit, as shown in Figure 1.22. In this case, output voltage is equal to the sum of the input voltage and the charge stored across the capacitor. The resulting output of clamper circuit is shown in Figure 1.23. It can be observed that by using this clamping circuit, the output of the circuit shifted with negative DC value without change in the peak-to-peak amplitude and it is expressed as equation (1.8)

$$V_{out} = -\left(V_C + V_{in}\right) \tag{1.8}$$

where V_C is the voltage stored with the capacitor in the positive half cycle of input.

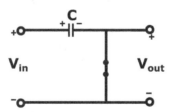

FIGURE 1.21 Clamper circuit during positive cycle of input.

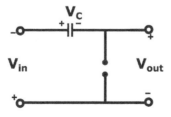

FIGURE 1.22 Clamper circuit during negative cycle of input.

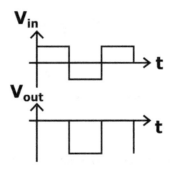

FIGURE 1.23 Clamper output showing negative shifting.

Similarly, positive clamper and its characteristics can be observed from Figure 1.24. The other types of clamper circuits also exists where we can define the voltage level of clamping by changing the direction of the diode, and the DC source of different clamper circuits can be generated with distinct DC value shifting.

1.3 BIPOLAR JUNCTION TRANSISTOR

Bipolar junction transistor (BJT) is a first and important transistor invention before the field-effect transistor (FET). It is a bipolar device, which means that device exhibits current conduction using both electrons and holes. Here, we will describe the structure and characteristics of the BJT.

1.3.1 SYMBOL AND PHYSICAL STRUCTURE

BJT is available in two different types: *NPN* and *PNP* BJTs. The BJT is a three-terminal device. The terminals are labelled as the base, the emitter, and the collector. The symbols are shown in Figure 1.25.

The majority charge carriers for *NPN* and *PNP* transistor are electrons and holes, respectively. Here, we are considering an *NPN* transistor to explain the detail about the structure and characteristics. Figure 1.26 shows the *NPN* BJT having emitter, base, and collector regions. Of these regions, the collector has a large area and the base has a low area. In BJT, the emitter, base, and collector, are heavily doped, moderately doped, and lightly doped regions, respectively [4]. The BJT has two junctions known as emitter-base junction and base-collector junction. Based on the biasing

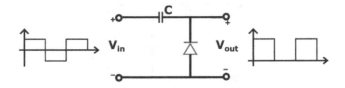

FIGURE 1.24 Clamper output showing positive shifting.

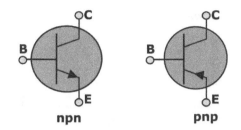

FIGURE 1.25   Symbols of BJT.

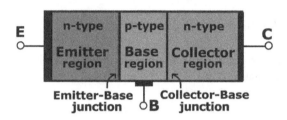

FIGURE 1.26   Physical structure of NPN BJT.

conditions (reverse and forward) of these junctions, BJT can be operated in different regions, as summarized in Table 1.3.

1.3.1.1   Operation and Several Current Components

When both emitter-base and collector-base junctions are in reverse bias, the transistor operates in the cut-off region wherein collector, emitter, and base current will be zero. The operation of BJT in the active region is illustrated by using *NPN* transistor structure, as shown in Figure 1.27. Two biasing voltages are applied to make emitter-base junction forward biased and collector-base junction reverse biased.

Due to the forward biasing of the emitter-base junction, emitter current flows out from the emitter. Base current will be low compared to that of the emitter because of lower doping concentrations of the base. When electrons move from the emitter to the base, the majority of charge carriers will be attracted by the collector due to the high positive voltage applied to the collector. This contributes collector current

TABLE 1.3
Different Operating Modes of BJT

Operating regions	Emitter-Base Junction	Collector-Base Junction
Cutoff	Reverse	Reverse
Active	Forward	Reverse
Reverse active	Reverse	Forward
Saturation	Forward	Forward

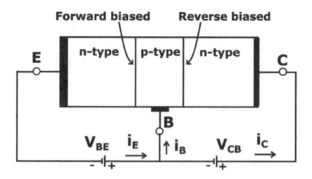

FIGURE 1.27   NPN BJT connecting in active mode.

in BJT. Since the current enters from transistor must leave, the emitter current (i_E) can be related as equation (1.9).

$$i_E = i_C + i_B \tag{1.9}$$

We can express collector current as

$$i_C = I_s e^{V_{BE}/V_T}, \tag{1.10}$$

and base current

$$i_B = \frac{I_S}{\beta} e^{V_{BE}/V_T} \tag{1.11}$$

From both equations (1.10) and (1.11)

$$i_C = \beta i_B \tag{1.12}$$

where β is called common emitter current gain

$$i_C = \frac{\beta}{(1+\beta)} i_E \tag{1.13}$$

where constant α (common base current gain) is related to β by

$$\alpha = \frac{\beta}{(1+\beta)} \tag{1.14}$$

1.3.2   BJT CONFIGURATIONS

With three terminals, BJT can be connected in three different configurations making one terminal common to input and output. Each configuration provides

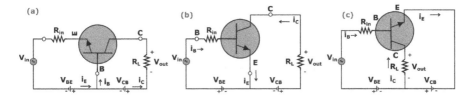

FIGURE 1.28   (a) Common base, (b) common emitter, and (c) common collector configurations of BJT.

different characteristics and suits to distinct applications. The three configurations include:

- Common base configuration
- Common emitter configuration
- Common collector configuration

1.3.2.1   Common Base (CB) Configuration

In this configuration, the base terminal is common to both the input and output. Input signal is applied between the emitter and the base, and the output is measured from the collector and the base, as shown in Figure 1.28(a). The input current through the emitter is larger than the collector current; due to this, the output at the collector is lesser compared to the emitter. As a result, the CB configuration attenuates the voltage signal. However, this topology shows a current gain of unity.

1.3.2.2   Common Emitter (CE) Configuration

In this configuration, the emitter terminal is common to both input and output. The input signal is applied between base and emitter, and the output is measured from the collector and the emitter, as shown in Figure 1.28(b). The input current through the base is small and the output collector current is very high. Due to this, CE configuration is used as an amplifier with high voltage gain.

1.3.2.3   Common Collector (CC) Configuration

In this configuration, collector terminal is common to both input and output. The input signal is applied between the base and the collector; the output is measured from the emitter and the collector, as shown in Figure 1.28(**c**). This type of configuration is commonly used in voltage follower or buffer applications. The summary of the characteristics of three configurations is given in **Table 1.4**. From this, it can be observed that CC configuration of BJT provides low power gain and hence CE configuration is not widely used to design a BJT-based amplifier.

1.3.2.4   BJT in CE Configuration: Operation and *I–V* Characteristic

With three terminals, BJT operates in different regions of operations. This can be explained using *I–V* characteristics; here, we have considered CE configuration of BJT. Most importantly, BJT operation can be explained by using two characteristics:

- Input characteristics (drawn between V_{BE}, I_B)
- Output characteristics (drawn between V_{CE}, I_C)

TABLE 1.4

Comparison of BJT Circuit Configurations

Characteristics	Common Base (CB)	Common Emitter (CE)	Common Emitter(CC)
Input impedance	Low	Medium	High
Output impedance	Very high	High	Low
Phase angle	0°	180°	0°
Voltage gain	High	Medium	Low
Current gain	Low	Medium	High
Power gain	Low	Very High	Medium

1.3.2.4.1   Input Characteristics

Input characteristics can be drawn between the input current (I_B) and the input voltage (V_{BE}). By varying the V_{BE}, the input current will be recorded at constant V_{CE}. Figure 1.29 shows the input characteristics that resemble as *PN* diode since the base-emitter junction behaves as forward-biased *PN* diode.

1.3.2.4.2   Output Characteristics

Output characteristics can be drawn between the output current (I_C) and the output voltage (V_{CE}). By varying the V_{CE}, the output current will be recorded at constant V_{BE}. Figure 1.30 shows the output characteristic, which shows the different regions of operations by varying the value of V_{CE}.

1.3.3   Second-Order Effects

Various unusual effects occur in BJT, which are not explained in the conventional operation of transistor. These effects include the base-width modulation, high injection effects, temperature dependence, and breakdown mechanisms in BJTs.

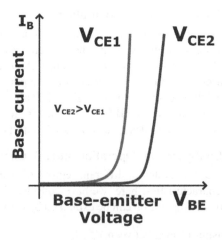

FIGURE 1.29   Input characteristics of CE configuration of BJT.

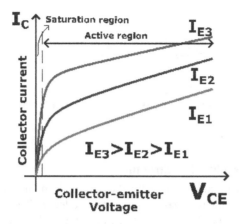

FIGURE 1.30 Output characteristics of CE configuration of BJT.

1.3.3.1 Base-Width Modulation

When voltage is applied to base-emitter and base-collector junction changes, the depletion layer of the transistor varies. Due to this, the collector current increases which can be referred as the *"Early Effect."*

1.3.3.2 Recombination in the Depletion Region

Collector current does not include the recombination current in the depletion region which will increase the overall value of the current.

1.3.3.3 Breakdown Mechanism in BJT

When reverse collector base voltage increases, the collector current increases. Due to this, avalanche multiplication happens that finally makes the device breakdown.

1.4 FIELD-EFFECT TRANSISTOR

The FET is a voltage-controlled device, which depends on the applied electric field. There are mainly two important transistors available with respect to the structure; they are junction field-effect transistor (JFET) and metal-oxide-semiconductor field-effect transistor (MOSFET). Here, we mainly discuss

- Principle of operation of JFET and their characteristics
- MOSFET structure and characteristics

1.4.1 JUNCTION FIELD-EFFECT TRANSISTOR (JFET)

JFET is a unipolar device and their operation depends on the flow of charge carriers. There are two types of JFETs depending on the channel type. They are:

- *N*-channel JFET
- *P*-channel JFET

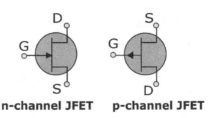

n-channel JFET   p-channel JFET

FIGURE 1.31   JFET symbols.

1.4.1.1   Symbol and Physical Structure

JFET is a three-terminal device, with the terminals named as gate, drain, and source, which are identical to base, emitter, and collector. The symbols of the both N-channel and P-channel JFETs are shown in Figure 1.31. The arrow mark on the gate terminal indicates the flow of current when gate and source are forward biased. The physical structure of N-channel and P-channel JFET is shown in Figure 1.32. An N-channel JFET consists of N-type silicon bar with P-type regions diffused on both sides. The two edges of the bar are treated as source and drain, and the P-type regions are connected together to form a gate terminal.

Source: It sources majority charge carriers into the channel.
Drain: Drain collects the charge carries from the channel.
Gate: Gate controls the flow of current by using the voltage applied to it.
Channel: It is the region between two gate regions and allows the flow of charge carries from source to drain.

1.4.1.2   Operation of JFET

Considering the N-channel JFET for explaining the operation, it is assumed that the two gate terminals are tied together. When gate-to-source voltage (V_{GS}) is 0, the gate regions become forward bias and current flows from the drain to source with positive drain to source voltage (V_{DS}), as shown in Figure 1.33**(a)**. When V_{GS} is becoming negative, the depletion region of gate channel increases that increase

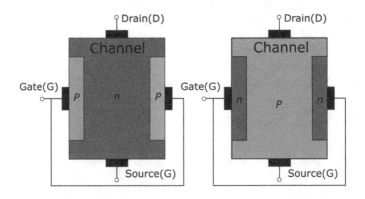

FIGURE 1.32   JFET physical configuration.

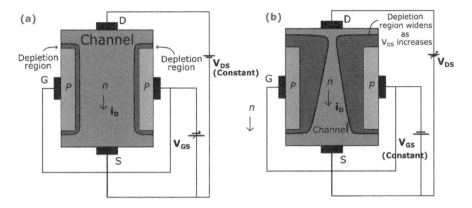

FIGURE 1.33 Behavior of JFET (a) with $V_{GS} = 0$ (b) with negative V_{GS} and increasing V_{DS}.

the channel resistances and hence the drain current reduces. When V_{GS} is becoming more negative, the depletion region occupies entire channel that makes the channel fully depleted. This value of V_{GS}, where channel gets depleted, is called pinch-off voltage (V_P).

Now, consider another case where V_{GS} is fixed at the constant value, which is lesser than V_P. When V_{DS} increases, the resultant reverse bias voltage varies by moving from source to drain. Consequently, the reverse bias is high at the drain end and the channel acquires a shape of tapered, as shown in Figure 1.33(**b**).

1.4.1.3 Current–Voltage Characteristics and Regions of Operation

The current-voltage characteristics for the above two explained conditions are shown in Figure 1.34. JFET can work in three regions of operation. Those include:

Cut-off region:

$$V_{GS} \leq V_P, \, i_D = 0 \tag{1.15}$$

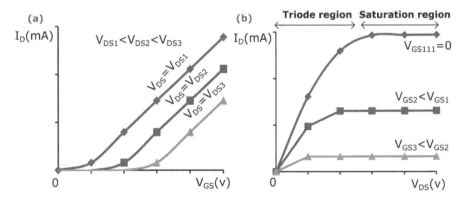

FIGURE 1.34 *N*-channel JFET (a) transfer characteristics (b) output characteristics.

Triode region:

$$V_P \leq V_{GS} \leq 0, V_{DS} \leq V_{GS} - V_P \tag{1.16}$$

$$i = I_{DSS}\left[2\left(1 - \frac{V_{GS}}{V_P}\right)^2\left(\frac{V_{DS}}{-V_P}\right) - \left(\frac{V_{DS}}{V_P}\right)^2\right] \tag{1.17}$$

Saturation region:

$$V_P \leq V_{GS} \leq 0, V_{DS} \geq V_{GS} - V_P \tag{1.18}$$

$$i = I_{DSS}\left(1 - \frac{V_{GS}}{V_P}\right)^2\left(1 + \lambda V_{DS}\right) \tag{1.19}$$

where λ is the inverse of early voltage, $\lambda = \frac{1}{V_A}$.

1.4.2 METAL-OXIDE-SEMICONDUCTOR FIELD-EFFECT TRANSISTOR

MOSFETs are named due to their architecture and show high input impedance compared to the JFET. Due to this reason, MOSFETs are widely used for IC design. MOSFETs are of two types [5]:

Depletion-mode MOSFETs: These are normally ON switches that require gate-to-source voltage to switch OFF the device.
Enhancement-mode MOSFETs: These are normally OFF switches that require gate-to-source voltage to switch ON the device.

Depletion-mode MOSFETs characteristics are almost similar to JFET, because of which depletion-mode MOSFETs are not studied much. Enhancement MOSFET is considered and its physical operation has been explained in detail.

1.4.2.1 Symbol and Device Structure

Enhancement MOSFETs are also classified as *N*-channel and *P*-channel MOSFETs. MOSFETs consist of four terminals: drain (D), source (S), gate (G), body (B). The symbols of both *N*-channel and *P*-channel enhancement MOSFETs are shown in Figure 1.35.

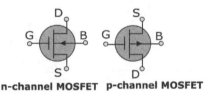

n-channel MOSFET   p-channel MOSFET

FIGURE 1.35   MOSFET symbols.

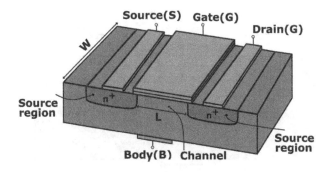

FIGURE 1.36 *N*-channel MOSFET physical structure.

Here, we consider *N*-channel MOSFET (NMOS) for reviewing its structure, as shown in Figure 1.36. The *N*-channel enhancement MOSFET fabricated on *P*-type substrate and two heavily doped *N*-type regions are formed to serve as source/drain. The *P*-type substrate also acts as the body of the MOSFET. A thin layer of silicon dioxide (SiO_2) is formed on the top of the substrate, between source and drain, which acts as an insulator. Metal layer is formed on the top of source, gate, drain, and body to have a contact with all the terminals.

1.4.2.2 Device Operation

With zero voltage applied to gate-to-source (V_{GS}) terminal, the two back-to-back connections of diodes exist between source body and drain body. As a result, no current flows between source and drain with the application of positive drain to source voltage (V_{DS}). With this biasing, transistor will be in cut-off region and zero current flows, as shown in Figure 1.37.

$$\text{When } V_{GS} = 0, \, i_D = 0 \text{ ; Cut-off region} \qquad (1.20)$$

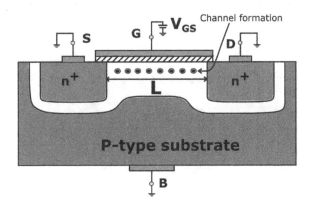

FIGURE 1.37 *N*-channel MOSFET with zero V_{DS}.

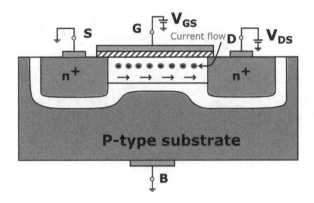

FIGURE 1.38   *N*-channel MOSFET with a small V_{DS} value.

With positive V_{GS} and V_{DS} applied, gate pushes the positive charge carriers down the body and the full of negative voltage will be formed between drain and source. The induced *N*-region forms a channel with an application of positive V_{DS}, as shown in Figure 1.38 and charge carriers flow between source and drain. The minimum value of V_{GS} that is required to form a channel is called threshold voltage (V_t).

With the small value of V_{DS}, the current varies linearly and increases with increase in drain voltage. This region is called triode region and in this region, the circuit works as a resistor. The drain current of MOSFET in triode region is modeled as

$$\text{When } V_{GS} \ge V_t, \ V_{DS} \le V_{GS} - V_t; \text{ Triode region} \qquad (1.21)$$

$$i_D = \mu_n C_{ox} \left(\frac{W}{L} \right) \left[\left(V_{GS} - V_t \right) V_{DS} - \frac{1}{2} V_{DS}^2 \right] \qquad (1.22)$$

With an increased V_{DS}, the channel obtains the tapered shape, as shown in Figure 1.39, and the channel depth at drain side reduces to zero. This effect is called pinch-off, and the current in this region becomes constant. The drain current thus saturates at this

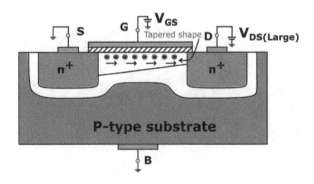

FIGURE 1.39   *N*-channel MOSFET with a large value of V_{DS}.

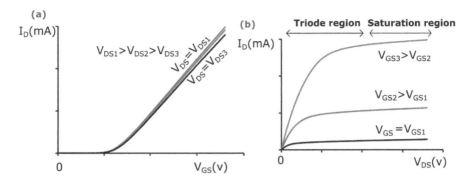

FIGURE 1.40 An *N*-channel MOSFET (a) transfer characteristics (b) output characteristics.

value and therefore this region is named as saturation region of MOSFET. The drain current of MOSFET in saturation region is modeled as

When $V_{GS} \geq V_t$, $V_{DS} \geq V_{GS} - V_t$, saturation region

$$i_D = \frac{1}{2}\mu_n C_{ox}\left(\frac{W}{L}\right)\left(V_{GS} - V_t\right)^2 \tag{1.23}$$

The transfer characteristics and output characteristics of an NMOS transistor are shown in Figure 1.40. Transfer characteristics are drawn between input voltage (V_{GS}) and the drain current; it shows that NMOS produces the drain current when V_{GS} is greater than the threshold voltage. The output characteristics are drawn between output voltage (V_{DS}) and drain current and the region of operations is indicated in Figure 1.40(b). As discussed in the operation of MOSFET, at small value of V_{DS}, the MOSFET operates in the triode region; when V_{DS} increases, it enters into the saturation region.

1.4.3 ADVANTAGES OF MOSFET OVER JFET

MOSFET shows several advantages compared to the JFET, as a result, MOSFETs have become popular in IC design.

- MOSFETs exhibit high input impedance compared to JFETs.
- MOSFETs provide high packaging density.
- MOSFETs exhibit ultra-low power consumption.
- MOSFETs can be explored for different applications due to their distinct characteristics.

1.5 EMERGING DEVICES BEYOND CMOS

Energy efficiency is considered to be one of the most critical design parameters for Internet of Things (IoT). In order to have an improved functionality and performance without compromising on battery life, there is a need to explore emerging technologies that can overcome the limitations of CMOS technology scaling and deliver greater energy efficiency. This discussion includes

- Issues with current CMOS technology scaling.
- Different emerging technologies with their characteristics.

1.5.1   ISSUES WITH CMOS TECHNOLOGY SCALING

Progress in the field of semiconductors has followed an exponential behavior that has come to be known as Moore's law. But will these exponential projections come to pass or will CMOS physical limits make them impossible? Several researchers have written about the current state and future prospects for Si MOSFETs and some of its limitations at lower channel nodes are mentioned below.

1.5.1.1   Velocity Saturation and Mobility Degradation

With more device scaling, the resultant effect of electric field increases and charge carriers in the channel exhibit high velocity. Consequently, there will be no linear relation between electric field and charge carrier velocity. This effect is called velocity saturation causing a reduced saturation current. This causes the reduction in switching speed of MOSFET.

1.5.1.2   Tunneling Current Through Gate Insulator

With excessive scaling, huge tunneling current flows through the gate insulator of the MOSFET device. For conventional insulator SiO_2, this tunneling limit has been increased and is becoming a bottleneck for ultra-low power applications. To reduce these, high dielectric materials are required as insulators.

1.5.1.3   High Field Effects

In scaling trends of CMOS, the supply voltages have not been scaled at the same rate as the length. To increase the device speed and performance, the supply voltage has not been scaled in proportion to the channel length. Due to this, there exists high electric field at channel of MOSFETs. At such high fields, several undesirable effects occur; one such effect is hot carrier effect.

1.5.1.4   Power Limitation

The supply voltage used for circuit has not been scaled as fast as the channel length of MOSFET. As the power consumption of MOSFETs is directly proportional to the square of the supply voltage used, the power consumption will dominate when the channel length becomes lower.

1.5.1.5   Material Limitation

Materials like silicon (Si), silicon dioxide (SiO_2), aluminum (Al), copper (Cu), and salicide are reached to their physical limits and cannot show the expected performance. For example, SiO_2 reliability degrades as it becomes thinner; in this regard, researchers need to identify new materials to support physical scaling limitations.

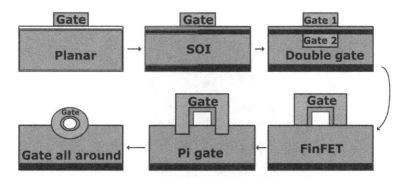

FIGURE 1.41 Evolution of device structure from single-gated planar to fully GAA NW MOSFETs.

However, the new materials may add fabrication difficulties and reliability issues; these parameters need to be considered while selecting the materials.

1.5.2 EMERGING NANOSCALE DEVICE TECHNOLOGIES

Recent emerging device technologies have been introduced and principle of operation of these devices is also discussed. Device characteristics have also demonstrated to show their benefits compared to existing CMOS.

1.5.2.1 Gate-All-Around (GAA) Nanowire (NW) MOSFET

To improve the resultant electric field, the multiple gate structures have been introduced. Figure 1.41 shows the evolution of multiple-gate transistors to increase the gate electrostatic control. Consequently, the GAA structure has become the most resistant to short-channel effects among all the emerging device structures for a given silicon body thickness [6].

1.5.2.2 Fin Field-Effect Transistor (FinFET)

A FinFET is a double-gate MOSFET where gate is wrapped around the channel, and the drain and source are formed as fins, as shown in Figure 1.42. This structure avoids the body in MOSFET, which reduces the leakage currents present in bulk MOSFET [7].

1.5.2.3 Carbon Nanotube FETs (CNTFETs)

Single-walled carbon nanotubes exhibit high conductivity and excellent carrier mobility due to their small diameter [8]. It has been experimentally demonstrated that these tubes can exhibit metallic or semiconducting characteristics depending on their chirality factor. Using semiconducting carbon nanotubes as channel element, CNTFETs have been demonstrated, as shown in Figure 1.43.

1.5.2.4 Tunnel FET (TFET)

Among several post-CMOS devices, TFETs have emerged as a promising device candidate for future low-energy electronic circuit design. TFET works with the

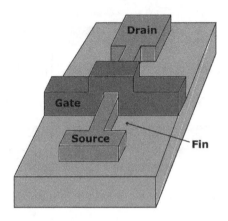

FIGURE 1.42   FinFET typical cross-sectional structures.

FIGURE 1.43   CNTFET structure.

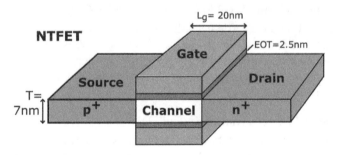

FIGURE 1.44   Tunnel FET device structure.

principle of band-to-band tunneling mechanism rather than thermionic emission of CMOS devices [9]. Consequently, it achieves high I_{ON}/I_{OFF} ratio and steep subthreshold swing (<60 mV/dec) at lower supply voltages, and TFET physical structure is shown in Figure 1.44.

1.6 SUMMARY

This chapter can be summarized as follows:

- This chapter discussed the semiconductors, and the different semiconducting devices invented and used in this field.

- First, we discussed the semiconductor types and properties of semiconductor. A comprehensive comparison was made among conductor, semiconductor, and insulators.
- Next, diode operation and different diode applications like rectifiers, gates, and clampers were discussed.
- Later, three-terminal devices like BJT, JFET, and MOSFET were discussed with their detailed structures and characteristics. Further, the benefits of MOS transistor against the JFET were highlighted.
- Finally, MOS transistor scaling issues were notified and emerging devices were briefly introduced.

1.7 MULTIPLE-CHOICE QUESTIONS

1. N-type silicon is obtained by doping silicon with impurity of
   a. Boron
   b. Aluminum
   c. Germanium
   d. Phosphorus

2. Identify the correct one.
   a. A silicon substrate doped heavily with boron is a P^+ substrate.
   b. A silicon substrate doped wafer lightly doped with boron is a P^+ substrate.
   c. A silicon substrate doped heavily with arsenic is a P^+ substrate.
   d. A silicon substrate doped lightly with arsenic is a P^+ substrate.

3. A BJT operates in the saturation region if
   a. Both the junctions are forward biased.
   b. Both junctions are reverse biased.
   c. Reverse-biased base–emitter junction and forward-biased base–collector junction is
   d. Forward-biased base–emitter junction and reverse-biased base–collector junction

4. The Early Effect in a bipolar junction transistor occurrs due to
   a. Large collector–base forward bias.
   b. Large emitter–base forward bias.
   c. Large collector–base reverse bias.
   d. Large emitter–base forward bias.

5. If the base width of BJT is doubled, which one of the following is true?
   a. Unity gain frequency will increase.
   b. Current gain will increase.
   c. Early voltage will increase.
   d. Emitter–base junction capacitance will increase.

6. In a uniformly doped BJT, the order of doping concentrations of BJT is
   a. Emitter = Base = Collector
   b. Emitter $\gg$ Base > Collector
   c. Emitter = Base < Collector
   d. Collector $\gg$ Base > Emitter

7. The drain of an N-channel MOSFET is connected to the gate. The threshold voltage of MOSFET is 1 V. If the drain current (I_D) is 1 mA for $V_{GS} = 2$ V, then for $V_{GS} = 3$ V, $I_D =$
   a. 2 mA
   b. 3 mA
   c. 9 mA
   d. 4 mA

8. MOSFET means?
   a. Metal oxide semiconductor field-effect transistor
   b. Metal oxide source field-effect transistor
   c. Metallic oxygen field-effect transistor
   d. Mercury oxide field-effect transistor

9. Which of the following statements is true?
   a. Input impedance of MOSFET is higher compared to BJT and JFET.
   b. Input impedance of BJT is higher compared to MOSFET and JFET.
   c. Input impedance of JFET is higher compared to BJT and MOSFET.
   d. Input impedances of MOSFET, BJT, and JFET are similar.

10. Which BJT configuration has a high voltage gain?
   a. Common Emitter
   b. Common Base
   c. Common Collector
   d. None

1.8   SMALL ANSWER QUESTIONS

1. What is doping?
2. What are P-type and N-type semiconductors? Give example.
3. Draw the symbol of PN diode.
4. Design a series clipper.
5. What are the properties of a BJT? Why is it named so?
6. What are the benefits of MOSFETs over JFETs?
7. Explain about CE configuration of BJT.
8. What is pinch-off in JFET?
9. Explain the different regions of operation of MOSFET.
10. What are the different non-silicon emerging transistors available?

1.9   LONG ANSWER QUESTIONS

1. Explain the conductor, semiconductor, and insulator with band diagram. Give examples.
2. Explain about doping of N-type and P-type materials.
3. Common emitter current gain β is 100 and base current is 15 μA. Calculate the emitter and collector current.
4. In a transistor, the base current and collector current are 70 μA and 1.85 mA, respectively. Calculate the value of α.
5. Explain the principle of operation of a Zener diode with characteristics.
6. A silicon substrate is uniformly doped with donor-type impurities with a concentration of $10^{17}/cm^3$. Electron mobility is 1200 cm^2/V-s and hole mobility is 400 cm^2/V-s, and charge of an electron is 1.6×10^{-19} C. Calculate electrical conductivity of semiconductor.
7. A MOSFET in saturation with gate-to-source voltage (V_{GS}) is 900 mV, has threshold voltage of 300 mV, and the drain current is observed to be 1 mA. What is the drain current for an applied V_{GS} of 1400 mV? (Neglect the channel width modulation.)
8. BJT common-base current gain α = 0.98 and reverse saturation current I_{CO} = 0.6 μA. BJT is connected in the common emitter mode and operated in the active region with a base drive current I_B = 20 μA. Calculate the collector current for this mode of operation.
9. Explain about clippers and clampers with examples.
10. Explain about different rectifiers and compare.
11. Explain the operation of BJT and detail about different region of operations.
12. Explain about different configurations of BJT. Compare them.
13. Explain the concept of JFET with characteristics.
14. Explain about the benefits of MOSFET and principle operation of MOSFET.
15. Explain about PN diode and characteristics.

REFERENCES

1. Schubert, T. F. Jr and Kim, E. M. 2015. Fundamentals of electronics: Book 2. *Synthesis Lectures on Digital Circuits and Systems* 10, no. 2: 1–366.
2. Fonstad, C. G., 1994. *Microelectronic Devices and Circuits*, McGraw-Hill, New York, pp. 504–524.
3. Kishore, K. L., 2008. *Electronic Devices and Circuits*, BS Publications, Hyderabad.
4. Sedra, A. S., Sedra, D. E. A. S., Smith, K. C. and Smith, E. K. C. 1998. *Microelectronic Circuits,* Oxford University Press, New York.
5. Boylestad, R. L. 2009. *Electronic Devices and Circuit Theory*, Pearson Education, India.
6. Liangchun, C. Y., Dunne, G. T., Matocha, K. S., Cheung, K. P., Suehle, J. S. and Sheng, K. 2010. Reliability issues of SiC MOSFETs: A technology for high-temperature environments. *IEEE Transactions on Device and Materials Reliability* 10, no. 4: 418–426.
7. Hisamoto, D., Lee, W. C., Kedzierski, J., Takeuchi, H., Asano, K., Kuo, C., Anderson, E., King, T. J., Bokor, J. and Hu, C. 2000. FinFET-a self-aligned double-gate MOSFET scalable to 20 nm. *IEEE Transactions on Electron Devices* 47, no. 12: 2320–2325.

8. Javey, A., Guo, J., Farmer, D. B., Wang, Q., Wang, D., Gordon, R. G., Lundstrom, M. and Dai, H. 2004. Carbon nanotube field-effect transistors with integrated ohmic contacts and high-κ gate dielectrics. *Nano Letters* 4, no. 3: 447–450.
9. Vallabhaneni, H., Japa, A., Shaik, S., Krishna, K. S. R. and Vaddi, R., 2014, March. Designing energy efficient logic gates with Hetero junction Tunnel fets at 20 nm. In *2014 2nd International Conference on Devices, Circuits and Systems (ICDCS)*, IEEE, pp. 1–5.

2 VLSI Scaling and Fabrication

2.1 INTRODUCTION TO VLSI SCALING

In the nascent stage of the MOS technology, electronic devices were complex and bulky, exhibiting degraded performance, higher maintenance cost, and poor lifetime [1]. These heavy and bulky electronic devices could not be moved easily from one place to another, presenting a challenge for the very-large-scale integration (VLSI) designers. Therefore, technology scaling using different photolithography processes was adopted by the designers decade by decade to reduce the size of electronic devices and improve their performance and reliability [2]. The progress in device scaling has proved to be a major contribution in the field of VLSI technology. The detailed information of VLSI scaling is explained later in Chapter 7.

2.1.1 HISTORY AND INTRODUCTION OF VLSI TECHNOLOGY

Twenty-first century has witnessed huge technical advancements. The application of the principles of electronics has led to a world where we are surrounded by several electronic devices and appliances. The history of the VLSI industry reveals the importance of electronic devices since the invention of the first point-contact transistor in 1947 [1]. Scientists John Bardeen, Walter Brattain, and William Shockley at Bell Labs succeeded in replacing the vacuum tube from their novel invention of the point-contact transistor, as shown in Figure 2.1(a) [3]. Later, the bipolar junction transistors (BJTs) were developed and they completely replaced the bulky, heavy, unreliable vacuum tubes soon after the invention of the transistor at the same Bell laboratory [4]. The adoption of BJTs completely transformed the electronics industry due to several advantages over the vacuum tube, such as more reliability, less noise, and reduced power consumption. Further, Jack Kilby succeeded in integrating the few transistors on a single silicon chip and fabricated the first integrated circuit (IC) at Texas Instruments just ten years after the invention of the first transistor, as depicted in Figure 2.1(b) [5].

The functionality of BJT limits its utilization due to the large leakage current at the base terminal. Therefore, the metal oxide semiconductor field-effect transistor (MOSFET) primarily became the first priority for the VLSI industry and was used for its advantages over the BJT by dissipating zero control current during the idle scenario. In 1963, the first logic gate was built using MOSFET technology by Frank Wanlass at Fairchild, and the logic gates consisted of p-channel metal oxide semiconductor (PMOS) and n-channel metal oxide semiconductor (NMOS) transistors known as complementary metal oxide semiconductor (CMOS) [5]. Currently, the

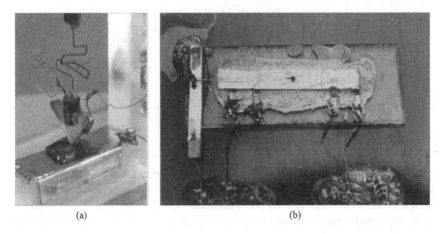

(a)   (b)

FIGURE 2.1 (a) First point-contact transistor at Bell Lab [3]. (b) Integration of a few transistors into the first integrated circuit [4].

CMOS is widely adopted in the electronics industry by the VLSI designers due to its ultra-low standby power consumption.

2.1.2 VLSI Design Concept

The concept of VLSI design is a complex process and designing an IC from the physical layout is carried out and processed with the hierarchal design process. Industry standard design tools such as *electronic design automation* (EDA) and *computer-aided design* (CAD) are widely used in order to design and test the circuit level, gate level, and layout in a fully or semiautomated manner before the actual fabrication [6–8]. VLSI designers implement their thoughts based on the required applications in the form of IC chips using the VLSI design flow, as shown in Figure 2.2 [9].

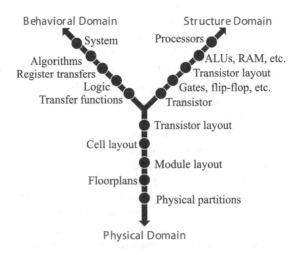

FIGURE 2.2 Design concept of an IC using Y-chart.

The designing of semiconductor devices is broadly followed by considering the Gajski-Kuhn Y-chart in the VLSI industry. It is divided into three categories as depicted in Figure 2.2: behavioral, structural, and physical domains. Moreover, the methodology of a VLSI design is broadly categorized as front-end and back-end designs. The front-end design primarily includes the design of digital circuits using HDLs such as Verilog, Verilog hardware description language (VHDL), and System Verilog. The back-end design comprises CMOS library design, physical design, fault testing, fault identification, and its several characterizations.

2.1.3 Moore's Law

The journey of the VLSI industry started in parallel with the concept given by Gorden Moore in 1965. After several observations for a number of transistors on a single chip with the technology enhancement, Moore [10] found that the number of transistors double in every 18 to 24 months from the journey of Intel microprocessor 4004 (Figure 2.3) [11]. With the evolution of technology, transistor count has increased, and die area has reduced. Later, Robert Noyce and Gorden Moore cofounded the well-known Intel Corporation, which went on to become, and still is, the world's largest chip manufacturer.

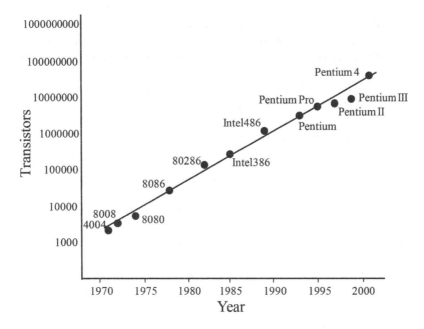

FIGURE 2.3   The growth of transistors on a single chip (Moore's law) [10, 11].

TABLE 2.1

Integration of Transistor on a Single Chip [12]

Level of Integration	Year	Number of Transistors on a Single IC Chip
Small-scale integration (SSI)	1950	Less than 100
Medium-scale integration (MSI)	1960	Up to 1000
Large-scale integration (LSI)	1970	Up to 10000
Very-large-scale integration (VLSI)	1980	Up to 1000000
Ultra-large-scale integration (ULSI)	1990	Up to 10000000
Global-scale integration (GSI)	2000	Higher than 10000000

2.1.4 SCALE OF INTEGRATION

The ICs can be categorized based on the number of transistors on a single chip. It can be categorized as small-scale integration (SSI), medium-scale integration (MSI), large-scale integration (LSI), very-large-scale integration (VLSI), ultra-large-scale integration (ULSI), and global-scale integration (GSI), as summarized in Table 2.1.

2.1.5 TYPES OF VLSI CHIPS (ANALOG AND DIGITAL)

The VLSI chips are broadly categorized into analog- and digital-based circuit implementation. The functionality of a digital IC depends on the binary level of inputs, that is, 0's and 1's. The analog ICs basically work by processing the continuous signal and are used for amplification, filtering, modulation, etc. The design process of a digital IC is mostly automated and transistor on a chip consumes less power and supply in comparison to the analog IC. The hardware description language (HDL) is used to describe the digital circuits, while analog cannot be described using HDL, and the analog modules need to be separated from the entire design process due to separate requirements of the ground terminals. In the case of analog VLSI, the impact of noise is more severe due to direct injection into the real system while testing the circuits. In a digital VLSI circuit, it is easy to identify the faults and has a lower impact on noise.

2.1.6 LAYOUT, MICRON, AND LAMBDA RULES

The design of the layout must follow some predefined set of rules before proceeding for an actual fabrication that is required in several masking processes. This set of rules is established to specify the geometrical structure of an object. The layout of an object must follow the minimum width and spacing between it to maintain the geometrical area. These design rules are the bridge between the circuit-level designer and the IC fabrication engineer. The circuit designer always prefers to have a smaller design with efficient power utilization and high performance along with more packing density. However, the IC fabrication engineer is concerned only about the high yield of the process. The design rule can be specified in the following two forms:

Lambda (λ) rules: The lambda design rules are used to define all the dimensions of a geometrical feature size, as shown in Figure 2.4. The maximum

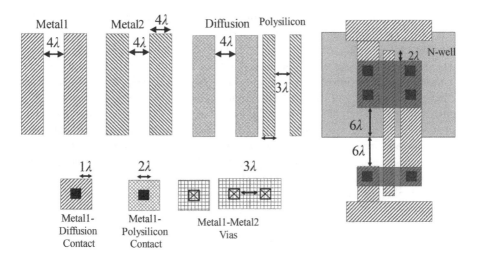

FIGURE 2.4   Dimensions as per the λ design rules.

distance of λ units is acceptable to the feature size of the fabrication process at a particular technology [13]. The feature size of a process is set to be 2λ that is widely used in the academic purpose.

Micron (μ) rules: The micron is mainly used in the foundries in the standard industry. The designer uses the absolute dimension by obeying the micron rules. It exhibits a 50% reduction in the design rules as compared to the lambda rules. This can differ from technology to technology, process by process, and company to company.

2.2   VLSI FABRICATION PROCESS

In this section, the detailed aspects of the process and manufacturing involved in the actual fabrication techniques are discussed. The importance of this section is to understand the consequences and basic knowledge behind the actual fabrication process.

2.2.1   PURIFICATION, CRYSTAL GROWTH, AND WAFER PROCESSING (CZ AND FZ PROCESS)

The detailed process related to purification and wafer processing is described as follows:

2.2.1.1   Introduction

The most naturally available semiconductor materials – such as silicon, chemical name Si, atomic number 14, and relative atomic mass 28.0855 (composite of silica and silicates) – are used as base materials due to their inherent properties in the semiconductor industry [14]. Most of the electronic devices are based on silicon that covers almost 95% of the electronics industry to date. Since the 1950s, the germanium was used as a semiconductor material, but several researchers reported the incompatibility with some of the applications due to high junction leakage current. The higher leakage current is due to the narrow bandgap of 0.66 eV of germanium. This led to the adoption of new semiconductor materials like silicon, having a higher energy bandgap of 1.170 eV and more stability compared to the germanium. The silicon wafer can be obtained from the natural material available in the pure form of sand via silicon dioxide SiO_2 known as *quartzite* [15]. In order to form pure Si, a high temperature is provided to the container that contains a mixture of SiO_2 and various composites of carbon such as coal, coke, and even wooden chips. This leads to the separation of SiO_2 bonding after reacting with carbon by producing carbon monoxide and Si at reduced temperature. The chemical reaction can be presented as

$$SiO_2 + 2C \rightarrow Si + 2CO\uparrow \tag{2.1}$$

The reaction involved in this process produces about 98 to 99% pure polycrystalline silicon or sometimes called crude silicon or metallurgical-grade silicon (MGS).

2.2.1.2 Electronic Grade Silicon

The Electronic Grade Silicon (EGS) is the raw material used to produce the single crystal silicon and a highly purified polycrystalline material. Even though the MGS exhibits less impurities in the order of parts per million (ppm), it is still required to be purified further such that the level of impurities is in the order of parts per billion (ppb) [16]. In order to obtain highly purified EGS from MGS, several reactions need to be accomplished through the chemical purification process. The following steps need to be performed:

1. The production of MSG inside the submerged electrode arc furnace shown in Figure 2.5.
2. The conversion of MGS into a volatile silicon compound such as monosilane (SiH_4), dichlorosilane (SiH_2Cl_2), and trichlorosilane ($SiHCl_3$) followed by distillation in order to purify the MGS.
3. Finally, the highly purified EGS can be obtained after decomposition.

The expression given below demonstrates the actual reaction that takes place inside the submerged electrode arc furnace.

$$SiC \text{ (solid)} + SiO_2 \text{ (solid)} \rightarrow Si \text{ (liquid)} + SiO \text{ (gas)} + CO \text{ (gas)} \tag{2.2}$$

The MGS used in this process is not suitable for metal alloys used in manufacturing solid-state devices. Therefore, an additional processing step is required in order to synthesize trichlorosilane ($SiHCl_3$) that is obtained by reacting anhydrous hydrogen chloride with silicon at 300°C, as shown below:

$$Si \text{ (solid)} + 3HCl \text{ (gas)} \rightarrow SiHCl_3 \text{ (gas)} + H2 \text{ (gas)} + Heat \tag{2.3}$$

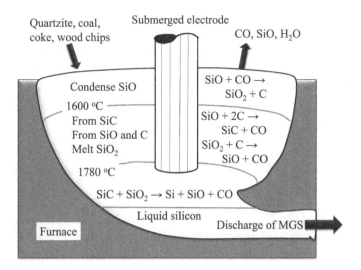

FIGURE 2.5 Schematic of electrode arc furnace to produce MGS [16].

Afterward, the chemical vapor deposition (CVD) process is performed to further purify the $SiHCl_3$ to form EGS.

$$2SiHCl_3 \text{ (gas)} + 2H_2 \text{ (gas)} \rightarrow 2Si \text{ (solid)} + 6HCl \text{ (gas)} \qquad (2.4)$$

An alternate approach to form EGS is more effective in terms of lower cost and less harmful reaction by products while using silane instead of trichlorosilane [17]. The reaction can be obtained as

$$SiH_4 \text{ (gas)} + \text{Heat} \rightarrow Si \text{ (solid)} + 2H_2 \text{ (gas)} \qquad (2.5)$$

2.2.1.3 Czochralski Crystal Growing

In general, the semiconductor-grade devices cannot be directly fabricated using polycrystalline silicon; therefore, the first step required in the wafer manufacturing process is the formation of the ingot or large single crystal silicon (i.e., monocrystal). Most common methodologies such as Czochralski (CZ) and the Float Zone (FZ) are used to obtain the silicon single crystal ingots.

1. **Czochralski method:** In this method, a feed material (polycrystalline silicon) is used as the raw material that is melted in a high-temperature furnace of 1400°C with a small doping amounts of As, B, P, and Sb. This mixture is melted in a highly pure quartz crucible in the presence of inert gas of argon. After melting the feed material into a cylindrical shape crucible, the seed rod is dipped from the top and a small portion of the seed is dipped into the melted feed [18]. Further, the dipped seed is slowly lifted while rotating in a circular motion and maintaining the furnace temperature. A new crystal shape is formed at the interface of the melted feed and the seed rod with a

controlled diameter by adjusting the temperature, pulling rate, and rotating speed. This methodology is important for the production of defect-free silicon by growing the small diameter of crystal in order to reduce the density of crystallographic defects. The arrangement of a Czochralski puller is shown in Figure 2.6.

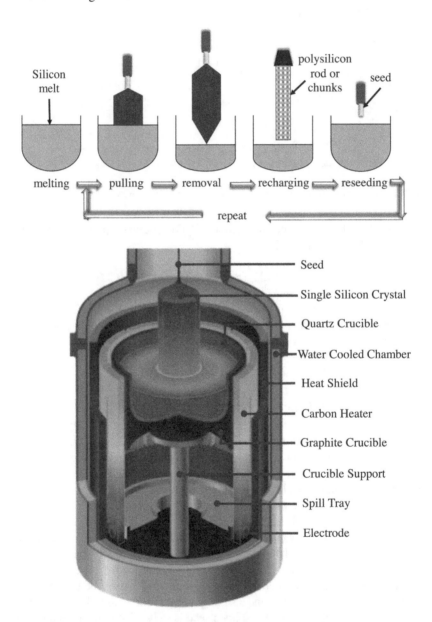

FIGURE 2.6   A schematic arrangement of a Czochralski puller [19].

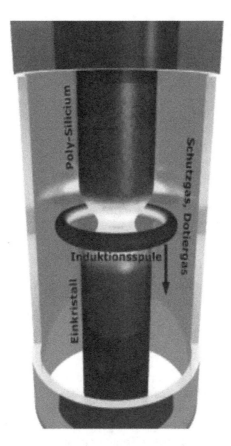

FIGURE 2.7   An experimental arrangement for the Float Zone process [19].

2. **Float Zone (FZ):** This is an alternate process of obtaining the highly pure silicon crystal as grown by the Czochralski methodology. The schematic diagram of the FZ is demonstrated in Figure 2.7, wherein the production takes place under vacuum or in an inert gas environment. Initially, the highly purified polycrystalline feed rod and a monocrystalline seed rod are placed face to face vertically on top of each other. The high temperature of the chamber (molten zone) is maintained such that the polycrystalline feed rod partially melts at the molten zone and the seed rod is brought up to make the contact with the drop of melt formed at the tip of the poly-crystalline silicon rod and simultaneously the material is purified [19]. Moreover, the purified material is passed through the necking process before it reaches to its desired diameter in order to establish a defect-free crystal. The comparative analysis is summarized based on the character-istics in Table 2.2.

TABLE 2.2

Comparison of CZ and FZ Characteristics

Characteristics	CZ	FZ
Grown speed (mm)	1–2	3–5
Crucible	Yes	No
Consumable material cost	High	Low
Heat or cool uptime	Long	Short
Axial resistivity	Poor	Good
Oxygen content (atom/cm³)	$>1 \times 10^{18}$	$<1 \times 10^{16}$
Carbon content (atom/cm³)	$>1 \times 10^{17}$	$<1 \times 10^{16}$
Metallic impurity content	High	Low
Production diameter	150–200	100–150

2.2.1.4   Silicon Shaping

Silicon shaping is an important process before the production of the wafer where they are used as the base material for IC manufacturing. The silicon shaping commonly requires three operations, namely, slicing, lapping, and etching. In order to perform these three operations, the monocrystal ingots obtained from either CZ or FZ are first required to maintain their uniform diameter with flat orientation [20, 21].

  i. **Slicing:** A diamond saw is primarily used to slice the wafer, wherein an inner-edge diamond saw is mostly used instead of an outer-edge diamond saw. This is because the use of inner-edge diamond saw results in lesser kerf of the blade edge and fewer irregularities in the final shape.
 ii. **Lapping:** In order to remove the rough surface and to maintain its symmetry after slicing with the inner-edge saw, the sliced wafer is then mechanically lapped. In this process a mixture of abrasive corundum (i.e., Al_2O_3) powder and water is applied on the wafer. Afterward, a wafer with a very smooth finish is produced by performing the four-way rotation on the wafer.
iii. **Etching:** Etching is the final stage in wafer processing in which the abrasive and damaged surface is repaired,, and microcracks are eliminated. Chemical solutions such as a mixture of nitric acid (HNO_3) and glacial acetic acid (CH_3COOH) are used to perform the etching process.

2.2.2   Oxidation

In this subsection, the readers can understand the first step of fabrication by understanding the thermal oxidation process. The growth mechanics and oxidation techniques can be further understood in order to have controlled oxides, and their important properties are as follows:

2.2.2.1   Introduction

In the fabrication of ICs, one of the most initial steps is to perform the oxidation process on top of the silicon wafer. High-quality oxide is required to form in a controlled and repetitive manner to maintain the characteristics. In order to achieve this, the thermal oxidation process can be performed using either dry oxidation or wet oxidation methodology based on the requirements. Therefore, an understanding of the oxidation process is primarily required for the designer; furthermore, there is a need to understand the electrical significance of the oxide.

2.2.2.2 Growth Mechanism

The development of high-quality silicon dioxide (SiO_2) made a reliable production of the commercial IC in the VLSI domain [22]. Of the several oxidation processes, such as thermal oxidation, electrochemical anodization, and plasma-enhanced CVD (PECVD), only thermal oxidation can provide the highest quality oxides that can work as excellent insulators. The friendly nature of silicon with oxygen rapidly forms an oxide layer under the influence of the oxidizing process. The chemical reaction for the dry and wet oxidation process can be obtained as below: [23]

$$\text{Dry oxidation process: Si (solid)} + O_2 \rightarrow SiO_2 \text{ (solid)} \qquad (2.6)$$

$$\text{Wet oxidation process: Si (solid)} + 2H_2O \rightarrow SiO_2 \text{ (solid)} + 2H_2 \text{ (gas)} \qquad (2.7)$$

Both the dry and wet oxidation processes are used to form a specific thickness of the oxide layer above the silicon wafer. Dry oxidation exhibits a good interface between Si and SiO_2 and forms a thin oxide layer in a device structure. Similarly, the higher growth rate of wet oxidation produces a thick layer of oxides that uses the water vapor. Figure 2.8 depicts the formation of the oxidation process involved before and after the reaction.

2.2.2.3 Oxidation Techniques and Systems

There are several techniques, such as peroxidation cleaning, dry oxidation, HCL dry oxidation, wet oxidation, and plasma oxidation, available based on the desired thickness and requirement of oxide properties in the fabrication process.

i. **Pre-oxidation cleaning:** In order to maintain the higher reliability and electrical characterization of an IC, the peroxidation process needs to be performed [24]. This process is primarily required such that any contaminant

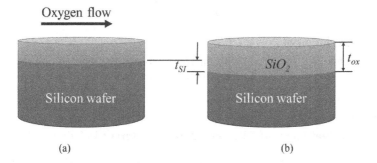

FIGURE 2.8 The oxidation process (a) before oxidation and (b) after oxidation, where t_{Si} and t_{ox} denote the thickness of silicon and silicon dioxide, respectively.

present in the form of organic and nonorganic particles in the silicon wafer can be eliminated using the chemical cleaning process. A high degree of cleaning must be followed before putting it into the high-temperature furnace to maintain the contaminant-free oxidation of silicon wafer.

ii. **Dry, HCL dry, and wet oxidation:** In this process, the microprocessor-based equipment is used directly to perform the dry and HCL dry oxidation to have a precise outcome. In this process, special care needs to be taken while handling the HCL and the amount of water vapor during the reaction with silicon wafer as it has a higher impact during the oxidation process.

iii. **High-pressure oxidation:** This type of oxidation process is mainly used to form an oxide layer at a relatively low temperature during run time and is comparable to typically high temperature, 1 atm conditions. The oxidation process using high-pressure drastically reduced the oxidation time [25]. The high-pressure oxidation also exhibits an excellent advantage for lower device dimensions at low operating temperatures by introducing suppressed defects and minimum lateral diffusion.

iv. **Plasma oxidation:** Unlike the high-pressure technique, plasma oxidation requires a much lower temperature to perform oxidation on top of the silicon wafer [26]. It also provides similar advantages as high-pressure oxidation along with the minimized movement of previous diffusions and protection from defect formation.

2.2.2.4   Redistribution of Dopants at Interface

At the time of the oxidation process, the silicon is either heavily or moderately doped with P-type or N-type depending on the case. In order to understand, we first define the term segregation coefficient that is nothing but the ratio of an equivalent concentration of dopant in the silicon to the equivalent concentration of the dopant in silicon dioxide [27]. The redistribution process mainly depends on the ratio of solubility of Si and SiO_2, where the redistribution of dopant will continue until the concentration at the interface becomes the same as the ratio of their solubility in both the materials [28]. The segregate coefficient can be denoted by k and expressed as equation (2.8). The k of different dopants in Si is summarized in Table 2.3.

$$k = \frac{\text{solubility in Si}}{\text{solubility in } SiO_2} \tag{2.8}$$

2.2.2.5   Oxidation of Polysilicon

In an IC technology, the conducting interface between the devices and the gates is established using polysilicon. Isolation between the multilayers of polysilicon can be

TABLE 2.3

Segregation Coefficient of Different Dopant Materials in Silicon

Dopant Materials	Antimony	Arsenic	Boron	Gallium	Phosphor
K	10	10	0.1–0.3	20	10

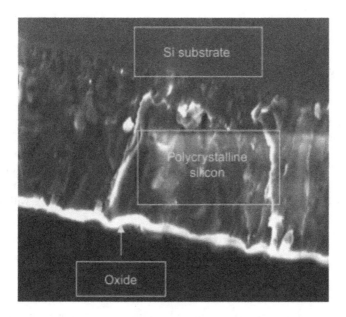

FIGURE 2.9   A thick polysilicon observed via SEM on 300 nm oxide [29].

provided using thermal oxidation that can protect to have interference between each other. Therefore, an understanding of the oxidation process is necessary for the designer such that the IC can last longer [29]. The oxide layer is grown on the boron-doped thick polysilicon layer, as shown in Figure 2.9, using a fast deposition process (40 nm/min) at 680°C.

2.2.3   EPITAXIAL DEPOSITION

In this subsection, readers can understand the importance of epitaxy on top of the single crystalline silicon in order to have a thin layer of ordered crystalline. Further, various types of epitaxy are also explained in the consequent subsections along with the evaluation of epitaxy.

2.2.3.1   Introduction

In the modern VLSI technology, the term epitaxy is derived from the Greek *epi*, means above, and *taxis*, an ordered manner that defines the creation of one or more thin crystalline layers in ordered orientation above the substrate [30]. Based on the composition of the substrate and the layer, it can be classified as homo-epitaxy and hetero-epitaxy. If the thin epitaxy layer is made up of the same material as the substrate, it is known as homo-epitaxy, whereas in hetero-epitaxy the epitaxy layer is of a material different from the substrate [31]. The epitaxial growth methodology is used to grow the highly purified single crystalline material that was first introduced by Royer in 1928. The electrical and optical properties can be precisely controlled using the techniques other than the bulk ones. There are several epitaxy methodologies available to form high-quality epitaxial growth, such as vapor-phase epitaxy (VPE) and molecular beam epitaxy (MBE).

2.2.3.2 Vapor-Phase Epitaxy

In VPE, the deposition of an atom can be performed using vapor such that the formation occurs between the interface of gaseous and solid matter. The growth of epitaxial silicon can be achieved using four silicon composites such as silicon tetrachloride ($SiCl_4$), dichlorosilane (SiH_2Cl_2), trichlorosilane ($SiHCl_3$), and silane (SiH_4) [32]. From the industry point of view, the most commonly used composite is $SiCl_4$ in the VPE from 1200 to 1250°C [33]. For example, the silicon tetrachloride reacts with hydrogen and produces $SiCl_2$ and HCl. Further, this $SiCl_2$ gets absorbed on the sample surface and forms silicon and $SiCl_4$. The chemical reaction with hydrogen gas is expressed as

$$SiCl_4 + H_2 \rightarrow SiCl_2 + 2HCl \qquad (2.9)$$

$$SiCl_2 + Si \rightarrow Si + SiCl_4 \qquad (2.10)$$

Similarly, chlorine can be used in place of chlorosilane when one or more hydrogen atoms are replaced.

$$SiCl_4 + H_2 \rightarrow SiHCl_3 + HCl \qquad (2.11)$$

$$SiHCl_3 + H_2 \rightarrow SiH_2Cl_2 + HCl \qquad (2.12)$$

The SiH_2Cl_2 obtained from the previous reaction can further react with hydrogen to form SiH_3Cl and HCl.

$$SiH_2Cl_2 + H_2 \rightarrow SiH_3Cl + HCl \qquad (2.13)$$

$$SiH_2Cl_2 \rightarrow SiCl_2 + H_2 \qquad (2.14)$$

$$SiCl_2 + H_2 \rightarrow Si + 2HCl \qquad (2.15)$$

This reaction is called surface catalysis that takes place only in the presence of the sample surface. This technique is used for mass production of electronic devices due to its low-cost, high-throughput, and advanced epitaxial structure. The process steps of the VPE are listed below and depicted in Figure 2.10.

i. Transfer of reactants to the region wherein the epitaxy needs to grow.
ii. Transfer of reactants to surface with a mixture of carrier gas and reactants.
iii. Absorption of reactants into the substrate that can handle surface processing such as a surface reaction and surface diffusion along with the site incorporation.
iv. The product desorption and transfer of the product into the main gas stream.
v. The final product away from the epitaxial growth region.

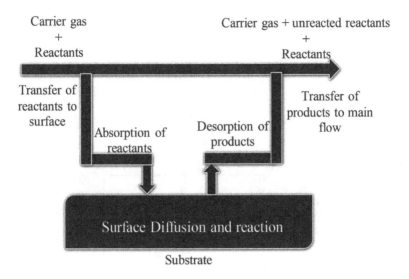

FIGURE 2.10 The steps followed to perform VPE.

2.2.3.3 Molecular Beam Epitaxy

In the late 1970s, J. R. Arthur and Alfred Y. Cho developed a technique to deposit a thin layer of single crystal that is widely used in the manufacturing of electronic devices [34]. This technique is used to fabricate the transistor; it also includes several components such as a diode, MOSFET, and lasers. This epitaxy methodology is performed using the evaporated technique instead of the CVD process. This process forms an atomic layer by layer thin crystal on top of the substrate depending on the nature of the crystal required to create. Figure 2.11 depicts the complete process involved to create a thin layer under ultra-high vacuum environments [35, 36]. First, a high temperature (approx. hundreds of degrees, 500–600°C or 900–1100°C) is provided to the substrate followed by a relatively precise beam of atoms that are fired one by one at the substrate using guns known as effusion cells. Each effusion cell contains a different kind of atom depending on the layer required on top of the substrate. Later, the condensation process is performed after the molecules reside on top of the substrate and slowly an ultra-thin film deposited. Since this process involves manipulation of different atoms on top of the substrate, it will be suitable for nanotechnology.

2.2.3.4 Silicon on Insulator

The concept of using an insulator on top of the silicon was introduced by C.W. Miller and P.H. Robinson in 1964. Later in 1979, the research group of Texas Instruments fabricated a silicon-on-insulator MOSFET and 3D IC with SOI CMOS in 1983 by the Fujitsu research group, respectively [37]. In 1995, IBM announced the first commercial use of SOI in CMOS technology. The use of an insulator on top of the silicon is motivated in order to reduce the impact of higher parasitic occurs

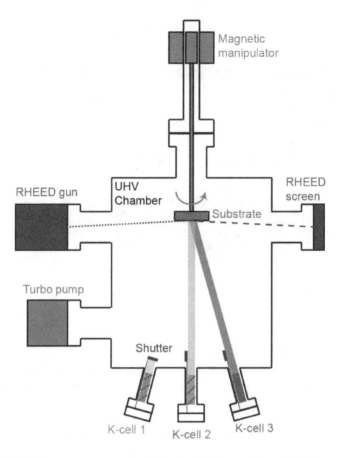

FIGURE 2.11 The process arrangement of molecular beam epitaxy [35, 36].

due to junction capacitance. Further scaling down of technology, the SOI primarily provides several advantages as follows:

i. It provides higher power efficiency because of the lower leakage current due to isolation.
ii. The power consumption improved due to isolation from the bulk silicon. It exhibits lower circuit parasitic.
iii. An equivalent V_{dd} provides higher performance.
iv. Several issues related to the antenna are resolved.
v. It utilizes the maximum area of the wafer.

2.2.4 LITHOGRAPHY

The detailed process of lithography is demonstrated in the following subsections:

2.2.4.1 Introduction

In 1796, lithography was first introduced by a German author, Alois Senefelder as a cheap method to transform the design into a base substrate. In general, it can be used to reprint the artwork or any print text on to another material or substrate [38]. Since 1960, most of the bulk production of books and magazines with colored images is done based on the offset lithography. After that, fabrication and mass production of ICs in the VLSI industry play a significant role. The conventional lithography was widely used but the cost of fabrication using conventional lithography was high and can be classified as photolithography and particle beam lithography. The different kinds of lithography methodology can be explained in subsequent subsections in a detailed manner.

2.2.4.2 Optical Lithography

A technique to transfer the image or layout of electronic components from the mask on to the surface of the semiconductor substrate with a photoresist thin material is known as optical lithography or photolithography. This process uses an ultraviolet light or visible light to form the pattern on photoresist material. There are two different kinds of photoresist material used in order to form a pattern such as negative and positive photoresist. Based on the solubility, one can use either positive or negative photoresist material [39].

Negative photoresist: The developer solution has less solubility, if the light exposed on the photoresist material forms negative images of the mask pattern on the silicon wafer surface.

Positive photoresist: The developer solution has higher solubility, if the light exposed on the photoresist material forms positive images of the mask pattern on the silicon wafer surface.

The process of photolithography involves several steps, as shown in Figure 2.12 to transfer the actual mask on to the wafer surface as follows [40]:

  i. First, the wafer surface layer needs to be cleaned followed by sufficient heating is provided in order to remove any moisture present on the wafer surface layer.
 ii. The spin coating of photoresist material (positive or negative resist) needs to apply on top of the wafer surface based on the applicability to develop the mask.
iii. The ultraviolet rays are used to develop the mask pattern based on the positive or negative resist on the wafer surface. The positive photoresist is most commonly used due to higher solubility during UV light exposure.
 iv. Finally, etching and stripping process are performed to remove the photoresist.

Advantages:

  i. It does not require any additional material and can etch the pattern or design to the silicon wafer using a single beam of UV light.
 ii. It is cost-effective and produces a pattern with high efficiency.
iii. The exact size and shape of the entire substrate can be controlled using this methodology.

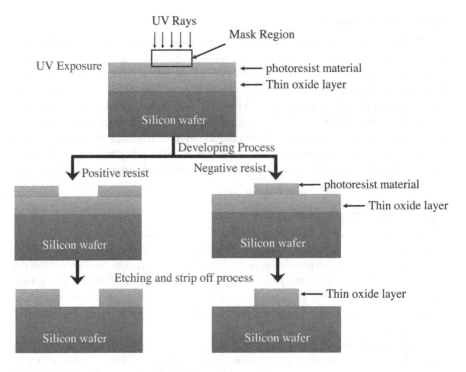

FIGURE 2.12   The steps involved in photolithography.

Disadvantages:

    i. In order to produce effective pattern, it requires a flat substrate.

    ii. It cannot produce an efficient pattern for an object that is not flat.

    iii. It requires a high degree of cleanness in order to avoid environmental hazards and any other contaminants.

2.2.4.3 Electron Beam Lithography

In the late 1960s, electron beam lithography (EBL) was the first developed technique that used the scanning electron microscope (SEM) integrated with control units. This process uses a common poly-methyl methacrylate (PMMA) polymer as an electron beam resist material till today [41]. In the nanotechnology regime, the EBL is used due to its tremendous advantages over conventional lithography. EBL exhibits several advantages such as higher resolution and precise pattern transform up to 20 nm; it reduces the impact on undesired pattern area, print complex pattern, flexible technique that can work for a variety of materials, and so on. The process flow of EBL to perform nanofabrication is presented in Figure 2.13 [41, 42].

The first step of the process involves cleaning the surface area of wafer followed by spin coating with electron beam resist material that is sensitive to the electron beam. Further, the light microscope (LM) and SEM together are used to provide control on developing the precise nanostructure. The LM and SEM can be used multiple times

FIGURE 2.13   The process flow of EBL technique.

in order to control and perform the exposure process. Following this, the proper alignment to the electron beam resist material needs to perform using "shuttle & find" to achieve an accurate pattern on top of the wafer surface. Furthermore, the nanostructure as per the design can be developed using a liquid developer. Finally, the dry or wet etching and the lift-off are together used to complete the nanofabrication using this methodology.

Advantages:
   i. It is capable of developing the complex pattern directly on the silicon wafer.
   ii. It does not have a diffraction problem.
   iii. It can develop the pattern with high resolution up to 20 nm as compared to the photolithography that can develop approximately 50 nm.
   iv. It is a flexible technique due to the direct development of the pattern on the wafer and does not require any mask. The pattern can be changed instantly based on the pattern using CAD from the system.

Disadvantages:
    i. It operates slower than optical lithography due to point-by-point exposure, which limits the speed.
    ii. The developing pattern on the wafer is expensive and complicated.
    iii. It exhibits forward and backscattering during the operation.

2.2.4.4   X-ray Lithography

The limitation of optical lithography provides a way to propose a new lithography technique known as X-ray lithography that can even work for short-wavelength. As the impact of diffraction reduces and resolution increases, the wavelength in optical lithography is also reduced. Further reduction in wavelength limits the performance of optical lithography due to the absorption of a signal on optical material. Therefore, the optical lithography becomes opaque but the transmission increases again for the X-ray region. In 1972, the X-ray lithography was the first technique that worked with an X-ray source [43]. Figure 2.14 depicts the various steps involved to perform X-ray lithography, wherein the substrate is coated with X-ray resist material. The X-ray source along with the mirror is used to provide controlled rays on top of the desired mask such that the required pattern can be developed on the silicon substrate.

Advantages:
    i. It can develop the pattern faster than the EBL technique.
    ii. It can provide a high resolution of approximately 0.5 μm.
    iii. It provides a reduced reflection, diffraction, and scattering effect at a short wavelength.
    iv. It provides a higher aspect ratio.
    v. The depth of focus problem can be solved using this methodology.

Disadvantages:
    i. Shadow printing is a major problem and has a lateral magnification error.
    ii. In order to operate, it requires brighter X-ray sources.
    iii. It is more sensitive toward the X-ray resist materials.
    iv. The production of the mask is very expensive.

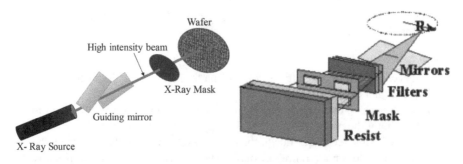

FIGURE 2.14   Schematic of X-ray lithography [43, 44].

2.2.5 POLYSILICON AND DIELECTRIC DEPOSITION

The process related to the deposition of polysilicon and dielectric materials is described below:

2.2.5.1 Deposition Process

In the VLSI design, deposition process uses a deposited materials to form a thin film that can provide electrical insulation between the metals, conducting region within a device, and protection from other environmental hazards. There are several methods to deposit thin film and the most commonly used materials are the polycrystalline silicon, silicon dioxide, and plasma deposited silicon nitride. The most widely used CVD methodologies are atmospheric-pressure CVD (APCVD), low-pressure CVD (LPCVD), and PECVD or plasma deposition [45–47].

Atmospheric pressure CVD: This method was first used in the VLSI microelectronics industry to deposit doped and undoped oxide material at atmospheric or normal pressure. At low temperature, the deposited oxide material exhibits low density and moderate coverage. The chemical reaction given below provides an advantage of higher wafer throughput when silane (SiH_4) reacts with the oxygen in the same chamber in order to form a thin film at about 430°C temperature.

$$SiH_4 + O_2 \rightarrow SiO_2 + 2H_2 \qquad (2.16)$$

The deposition technique is used based on the material deposited on top of the wafer, as depicted in Figure 2.15 [48, 49]. The wafer is kept on top of the moving belt through the rotter as a transport system and the processed gases are deposited from the top side. The deposition rates involved in this process are very fast in the order

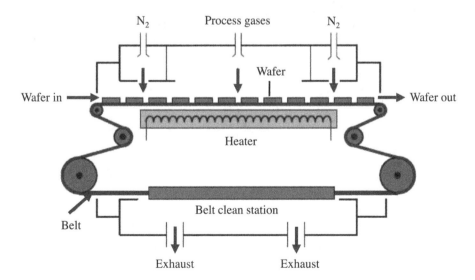

FIGURE 2.15 A mechanism to perform chemical vapor deposition based on atmospheric pressure [48].

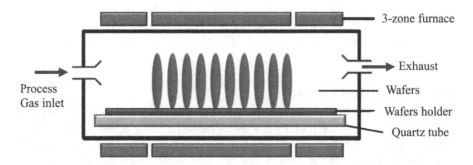

FIGURE 2.16   A general view of the low-pressure CVD process [49].

of 600 to 1000 nm/min. The whole process of AVCVD performs at a pressure of approximately 1 atm at 0°C. The moving wafer with the help of the transport system passes through a heater in order to accelerate the reaction process. Both the ends of the top chamber are used to supply nitrogen gas inside the chamber in order to protect the atmosphere from leaking the process gas outside. The same transport belt is continuously cleaned up to remove any contaminant present during the process.

LPCVD process: This system is used to produce various thin-film materials uniformly based on the requirement using low pressure, as depicted in Figure 2.16. In this process, the silicon wafer is put on top of the wafer holder and the processing gas is passed through the inlet such that the reaction is carried out with other gases inside the chamber. The reaction occurs inside the chamber in the presence of a hot or cold quartz tube reactor and provides high throughput and good thermal uniformity that results in a uniform thin film.

PECVD process: This process is used to deposit a thin film on top of the wafer at a low temperature between 100 and 400°C. In this process, thermal and cold plasma are used, wherein the particles in the gas and electrons are at the same temperature and cold plasma uses the energy of electron by changing its pressure. The schematic diagram of the process involved in the PECVD is demonstrated in Figure 2.17 that shows

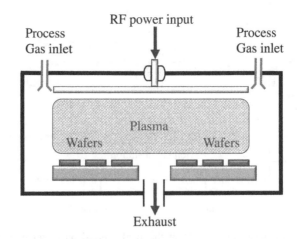

FIGURE 2.17   Schematic of PECVD process.

TABLE 2.4

Characteristics of the Various CVD Processes

Techniques	Advantages	Disadvantages
APCVD	Faster deposition, work at low temperature, and simplest reactor to process	Particle contaminant is present, throughput is low, and poor step coverage
LPCVD	The formation is uniform and provides excellent purity, large wafer capacity, conformal step size, and high throughput	Requires higher temperature to operate and the rate of deposition is low
PECVD	Rate of deposition is faster and required temperature is low, provides a good step coverage	Particle contaminant and chemical like H_2 present during the process

the wafer inside the chamber along with the gas mixture reacting at an extremely low temperature due to the presence of plasma particles [49]. The RF power input is provided from the top and the exhaust gases are pumped through the bottom side. The advantages and drawbacks of various CVD process are summarized in Table 2.4.

2.2.5.2 Polysilicon

In the modern VLSI industry, the silicon layer is categorized as epitaxial silicon and polycrystalline silicon (polysilicon). In CMOS technology, the gate electrode is often made of polysilicon along with most of the local interconnect as well due to higher sheet resistance as compared to a metal wire. In order to form polysilicon, the silane (SiH_4) is reacted with a high temperature using various CVD processes as per the requirement [50]. The chemical reaction is as follows:

$$SiH_4 \rightarrow Si + 2H_2 \qquad (2.17)$$

Polysilicon can be deposited on top of the silicon wafer using various methods such as diffusion, ion implantation, or in situ doping by following the CVD process. The polysilicon is heavily doped with impurities in order to use in MOS transistor as a gate electrode. The heavily doped polysilicon either with P-type or N-type is used to reduce the sheet resistance of the material.

2.2.5.3 Silicon Dioxide

In the VLSI industry, most of the fabrication processes utilize quartz glass or oxide that is known as silicon dioxide (SiO_2). The SiO_2 film can be deposited either by adding impurities or can be grown on top of the wafer. The gate oxide of the MOS transistor is made up of SiO_2, which acts as an excellent insulator and is compatible to a silicon wafer with a dielectric constant of 3.9. In order to develop silicon dioxide, silicon is reacted with oxygen in appropriate ambient surroundings. In this process, a thin dielectric film with an average growth rate of approximately 1.5 nm per hour with a thickness of 4.0 nm is produced. There are several CVD processes that can deposit the thin film of silicon dioxide on top of the silicon wafer and can be used in a variety of places at the time of fabrication of CMOS ICs.

2.2.5.4 Silicon Nitride

Silicon nitride (Si_3N_4) is used to prevent the circuit during fabrication from other contaminants present in the environment. It can be used for passivating silicon devices that act as a good barrier to the diffusion of water and sodium [51]. The dielectric constant of silicon nitride is relatively high at about 7.8 and is also used as a mask for the selective oxidation of silicon. Silicon nitride, also known as nitride and mostly deposited with silane (SiH_4) and ammonia (NH_3), reacts at a temperature of about 900°C [52, 53]. The chemical reaction can be observed as

$$3SiH_4 \text{ (gas)} + 4NH_3 \text{ (gas)} \rightarrow Si_3N_4 + 12H_2 \text{ (gas)} \qquad (2.18)$$

2.2.6 Diffusion

The detailed process of diffusion is described in the following subsections:

2.2.6.1 Introduction and Model of Diffusion

In a semiconductor, the impurity atoms are diffused in a controlled manner into the silicon wafer to alter the conductivity of the material. There are various procedures involved to introduce the dopant atoms into the silicon wafer. The dopants used in the diffusion process can be either P-type or N-type based on the impurity as a trivalent or a pentavalent. The most common trivalent P-type dopant is boron (B) and the pentavalent N-type dopant is arsenic (As) and phosphorus (P). The phenomenon of diffusion process can be observed based on the movement of atoms from higher concentration (higher energy) to lower concentration (lower energy) until the concentration becomes uniform. The process involved is called diffusion. The process can be easily understood based on the impurity atoms added on the silicon in order to change the conductivity. The diffusion rate of the impurity atom onto the silicon depends on the applied temperature [54]. The diffusion model can be elaborated in solid based on Figure 2.18, where Figure 2.18(a) demonstrates the vacancy diffusion mechanism by occupying the lattice site. However, Figure 2.18(b) depicts the movement of an interstitial atom around the lattice from one place to other without occupying the lattice site.

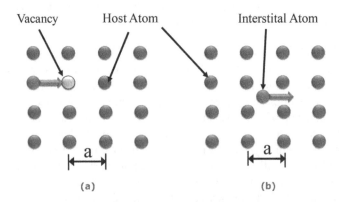

FIGURE 2.18   The diffusion of atom for (a) vacancy and (b) interstitial mechanism.

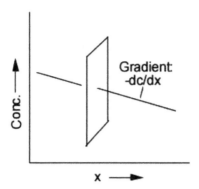

FIGURE 2.19 Schematic of Flick's first law of diffusion.

2.2.6.2 Flick's First Law of Diffusion

The flux of the particles flows from higher concentration to lower concentration with a magnitude that is proportional to the concentration gradient and can be expressed as [55]

$$J_{particles} = -D\frac{dC}{dx} \tag{2.19}$$

where $J_{particles}$, dC/dx, and D are denoted as the flux of the particles, the concentration gradient, and the diffusion coefficient or diffusivity, respectively. The negative sign on expression (2.19) defines the flow of particles from higher concentration to the lower concentration as depicted in Figure 2.19 that is known as Flick's first law. The diffusivity (D) can be given in the units of cm²/s or m²/s, C is in units of atoms cm⁻³.

2.2.6.3 Diffusion Factors

In general, the diffusion process primarily depends on the diffusivity (D) parameter. There are several factors involved that have a major impact on diffusivity such as the following:

Diffusing species: This phenomenon impacts the diffusion rate if the concentrations on both ends have a higher difference than the diffusion that takes place with a higher diffusive rate, whereas the diffusion rate becomes slower if the concentration on both sides is almost equal.

Temperature: The diffusion rate and the diffusivity have a higher impact due to variation on temperature. The diffusion rate increases with an increase in the temperature due to the energy gained by each particle to move faster from one place to another. Similarly, the impact becomes slower as temperature reduces due to the lower energy present in each particle and demonstrates a slower diffusion rate.

Size of molecules: The diffusion rate becomes faster with a smaller size of particles or molecules while the rate becomes slower with a larger size of particles. The particle present in the medium acts as a barrier to diffusion.

State of the matter: The diffusion becomes faster in gas molecules as compared to the liquid.

TABLE 2.5

Diffusivity of Elements in the Polysilicon Film

Impurity Elements	Frequency Factor D_0	Energy, E (eV)	Diffusivity, D	Temperature, T	References
As	8.6×10^4	3.9	2.4×10^{-14}	800	[56]
P	–	–	$(6.9-63) \times 10^{-13}$	1000	[57]
B	$(1.5-6) \times 10^4$	2.4–2.5	9×10^{-14}	800	[58]

Density of diffusion substrate: The region of higher density is responsible for lower diffusion due to the presence of a large number of particles.

2.2.6.4   Diffusivity in Polycrystalline Silicon and SiO_2

In modern VLSI technology, the use of polysilicon becomes more popular due to its major properties such as self-aligned as polysilicon gate and act as an intermediate conductor in a two-level structure. The diffusion sources, load resistor, the gate electrode, and local interconnect are realized using polysilicon for many important applications. The several doping elements such as boron, phosphorus, and arsenic are used to reduce the resistivity of polysilicon. The doping must be performed after the gate region is formed at low temperature in order to protect the polysilicon from diffusion in the gate oxide such that performance does not degrade. The diffusivity of several elements such as arsenic (As), boron (B), phosphorus (P), and antimony (Sb) in polysilicon is summarized in Table 2.5.

Similarly, silicon dioxide is used to protect the silicon wafer against the impurity elements before pre-deposition. Therefore, it is necessary to measure the quantitative value of diffusion of impurities in SiO_2 as illustrated in Table 2.6. The diffusivity of Group III and Group V elements strongly depends on their concentration, whereas Group III and V elements are formers in silicon dioxide such that it can lower the melting temperature of the oxide film.

2.2.7   Ion Implantation

Details of the steps involved in the ion implantation process are discussed in the following subsections.

TABLE 2.6

Diffusivity of an Element in Silicon Dioxide (SiO_2)

Impurity Elements	Frequency Factor D_0 (cm²/s)	Energy, E (eV)	Source and Ambient	References
As	3.7×10^{-2}	3.7	Ion implant, O_2	[59]
P	5.73×10^{-5}	2.20	P_2O_5 vapor, N_2	[60]
B	7.23×10^{-6}	2.38	B_2O_3 vapor, N_2+O_2	[60]
Sb	1.31×10^{16}	8.75	Sb_2O_5 vapor, O_2+N_2	[60]

2.2.7.1   Introduction

In the VLSI industry, the two different kinds of doping methods are used such as diffusion and ion implantation based on the requirement during the fabrication of electronic components, complicated structure, and devices on the silicon wafer. Using this approach, material properties can be modified by implanting high-energy ions on the silicon wafer. The energy of ions is mainly responsible for acquiring the changes in the material properties. In the ion implantation process, the beam of ion impurities is accelerated to kinetic energy and directed to the surface of the silicon wafer.

2.2.7.2   Range Theory

The basic phenomenon of ion implantation on a silicon wafer is covered in this subsection by defining the physics of ion collisions, range distribution, impact on a crystal lattice, damages, and recoil distributions.

Ion stopping power: Figure 2.20 demonstrates the behavior of implanted ions that strike the desired target while passing through several collisions with the host target until it reaches to the equilibrium state at some depth. The sum of nuclear and electron having energy per unit path length (l) is termed as the total stopping power P_s.

$$P_s = \left(\frac{dE}{dl}\right)_{electronic} + \left(\frac{dE}{dl}\right)_{nuclear} \tag{2.20}$$

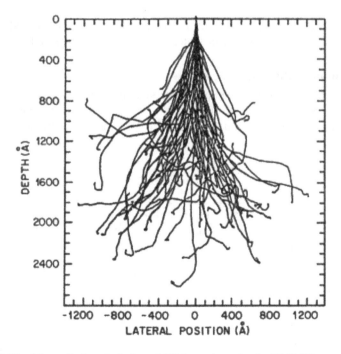

FIGURE 2.20   Monte Carlo calculation of 128 ion trajectories for 50 KeV boron implanted into silicon [61].

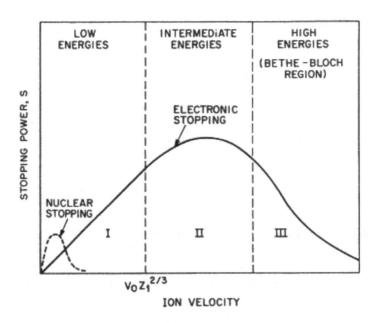

FIGURE 2.21 The ion stopping power *w.r.t.* ion velocity having nuclear and electronic stopping components at low, intermediate, and high energy [61].

Similarly, the ion stopping power *w.r.t.* ion velocity having nuclear and electronic stopping components at low, intermediate, and high energy is depicted in Figure 2.21.

2.2.7.3 Implantation Equipment

A general schematic of the ion implantation system is depicted in Figure 2.22 [62], whereas the ion source is used to ionize the desired impurity present in the form of gas. The ion beam generated from the source is separated with the help of mass

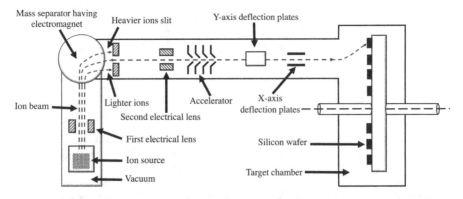

FIGURE 2.22 A general schematic of the ion implantation system using medium energy.

separator having electromagnet and the desired ions are directed through a narrow aperture precisely. Afterward, the resolved beams are passed through the second electrical lens that is used to focus the beam toward the accelerator. Finally, the deflection plates are used precisely such that the ions with required energy can strike the target and become implanted in the exposed area of the silicon wafer.

2.2.7.4   Annealing

In general, annealing is the process of heating a metal or alloy or semiconductor wafer at a desired temperature for a span of multiple times followed by a cooling chamber in order to change the mechanical and electrical properties. While performing ion implantation, the atoms are damaged due to ion bombarding, which results in the displacement of atoms. After ion implantation, the electrical property reduces and resistivity increases. Therefore, the annealing process is required to reactivate the dopant and repair the damages due to ion implantation.

Furnace annealing: In VLSI fabrication, heat treatment is a better choice to repair the damages and to activate the implanted dopant. The furnace annealing can be performed by following three processes namely, recovery, recrystallization, and grain growth. In order to complete this process, the silicon wafer is heated multiple times to soften the material at low temperature. This process is used to remove any linear defects or dislocation of atoms due to the ion implantation process. Then, recrystallization is done to make new grains strain-free and grow those that were removed during the recovery stage. At last, the growth of grain is performed only after the recrystallization has finished by applying appropriate heating as required.

2.2.8   Metallization

The final process of fabrication is metallization. The process of metallization primarily depends on the choice of metal and its deposition is as described below:

2.2.8.1   Choice of Metal

In the VLSI industry, interconnections are made by conducting films that are used to form interconnects, contacts, micro-mirror, fuse, and gates for CMOS technology in semiconductor devices. The selection of appropriate materials primarily depends on the work function of the material for the metallization process. Based on the property of the material, it can be used as a metallization for ICs. There are several materials such as aluminum (Al), tungsten (W), and copper (Cu), which have diverse properties for individual materials and are used based on different applications and requirements. Figure 2.23 demonstrates the different materials used for interconnection, barriers, and layers during metallization. The material used for metallization must have the following properties in order to achieve better performance:

i. The material should have low resistivity and should be easy to form on a silicon wafer during fabrication.
ii. It should be easy to perform the etch process for the pattern generation on the silicon wafer.

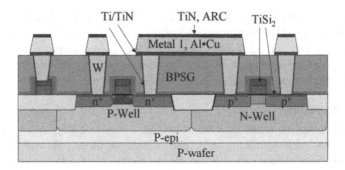

FIGURE 2.23 Cross-sectional view of a CMOS integrated circuit showing tungsten via and Al/Cu interconnection.

iii. It should maintain good adherence, low stress, and higher stability throughout the fabrication process.

iv. It should have a low Electromigration effect, low scattering, lesser defects, higher melting point, less contact resistance, good self-alignment, minimum junction penetration, and less environmental contaminant [63].

Apart from these properties, the material should have a lower RC time constant of the wire that varies with the dielectric material as expressed below [63]:

$$RC = \frac{\rho_{wire}}{d_{wire}} \frac{\varepsilon_{ox} L_{wire}^2}{t_{ox}} \qquad (2.21)$$

where ρ_{wire}, d_{wire}, L_{wire}, t_{ox}, and ε_{ox} are denoted as the resistivity of the material, the thickness of the wire, the length of wire, the thickness of the oxide, and the permittivity of the oxide, respectively. In most of the semiconductor industries, aluminum (Al) is widely used to form a metal layer using the metallization process since it offers a good conductivity and strong mechanical bonding with silicon. Further, Table 2.7 summarizes the applicability of different metals and alloys.

TABLE 2.7
Selection of Metals and Alloys for the Metallization on the Silicon Wafer

Metallization on Silicon Wafer	Choice of Metals and Alloys
Gates, interconnects, and contacts	Polysilicon, silicides, nitrides, carbides, borides, refractory metals, aluminum, etc.
Diffusion barrier	Nitrides, carbides, silicides, borides
Top level	Aluminum
Selectively formed	Tungsten, aluminum, and silicides

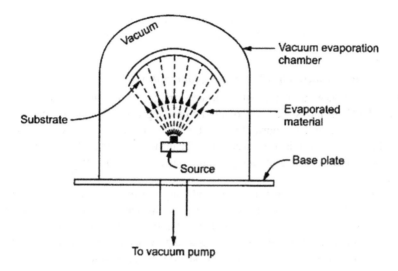

FIGURE 2.24   A general metallization process.

2.2.8.2   Metallization Process

In the VLSI fabrication process, the metallization process can take place in a vacuum-tight chamber by maintaining the chamber pressure in the range of 10^{-6} to 10^{-7}, known as the vacuum evaporation chamber as depicted in Figure 2.24 [64]. A bucket is used to keep the desired material that needs to evaporate whereas an electron gun is used as a source. Further, the electron gun is used to bombard onto the surface of the desired material and once the electron strikes on the surface, it gets stimulated and achieved enough energy to dissipate in the form of heat and vaporized. Later, the vaporized gas is condensed in order to form a thin film coating in the same chamber. Afterward, any excess aluminum present is removed by using the etching process, and finally, the developed thin film is used to form a desired interconnect pattern.

The metallization process is categorized into two important deposition processes such as CVD and physical vapor deposition. Further, the physical deposition process is categorized into two as evaporation and sputtering processes.

2.2.8.3   Metallization Problem

In the manufacturing and fabrication process, there are several challenges associated with the metallization process based on the material used for different applications such as

    i. The thickness of the metal film,
   ii. Uniformity of metal film,
  iii. The stress of metal film,
   iv. The reflectivity of the metal film,
    v. Sheet resistance and capacitance of metal film,
   vi. Deposition of metal films,

  vii. Metallurgical and chemical interaction with silicon wafer,
  viii. Electromigration and melting point related to the material used to form a thin film.

2.2.8.4   New Approaches Toward Metallization

In order to enhance the metallization process and better utilize it for different applications, the VLSI industry needs to adopt a new approach. Almost all the electronic devices and ICs play an important role in the new approach of the metallization process. The new approach is used to enhance the utilization and performance improvement based on different methodologies and they are as follows [65]:

  i. **Multilevel structures:** This methodology is used to minimize the resistance of interconnect and to enhance the utilization of area footprint, as shown in Figure 2.25 [65]. The deposition of thin film as a metallic interconnect in the multilevel stage is possible by using the dielectric material between each level to make them isolated from each other.
  ii. **Epitaxial metals, three-dimensional devices, and heterostructures:** In recent technology, the epitaxial process is a suitable approach to form the contact metallization in three-dimensional devices and hetero-structures. Using epitaxial silicides and aluminum, it is possible to form defect-free grain boundary and dislocation-free films that eliminate the issues related to the metallurgical and chemical interaction with silicon wafer. The stability and survival toward the electromigration increase by using the epitaxial scheme at a higher temperature.

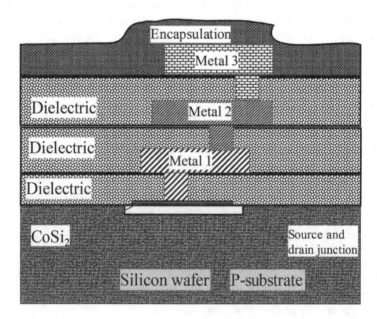

FIGURE 2.25 A schematic of multilevel structure having metallic lines separated with dielectric material.

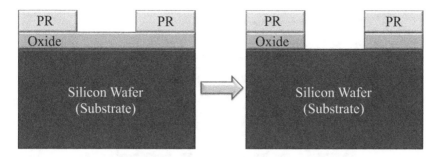

FIGURE 2.26 An etching process to remove unwanted material from the silicon wafer.

2.2.9 ETCHING PROCESS

The fabrication of the semiconductor IC is passed through several critical processes, whereas the silicon wafer is completely covered with a single- or multilayer of silicon dioxide, silicon nitride, or other metals. Therefore, the etching process is performed such that any unwanted material produced during the process can be removed as depicted in Figure 2.26. Primarily, there are two processes involved to remove the selected layers or multilayers from the silicon wafer: (i) dry or plasma etching (plasma-based etchant) and (ii) wet etching (liquid-based etchant).

2.2.9.1 Dry or Plasma Etching

In the VLSI industry, the plasma particles are used to remove unwanted materials from the silicon wafer. It is a special kind of ionized gas that consists of positive and negative charge in equal quantity. In order to perform dry etching, plasma can be generated by striking the high radio frequency to a gas molecule that will further break down the gas molecules into ions, free radicals, electrons, phonons, etc.

The mechanism behind plasma etching is the same as wet etching, where the gas molecules (such as Ar, CF_4, and O_2) are passed inside the chamber as depicted in Figure 2.27 [66]. Further, it will break down into ions, free radians, and electrons in order to form a plasma by imping the strong RF frequency. Afterward, the surface diffusion takes place once the plasma comes across the selected surface. Finally, based on the chemical reaction, the unwanted materials are dissolved or removed and any other gases are taken out using gas exhaust to form the final patterned product.

Advantages and disadvantages:

 i. Dry etching is cost-effective and has automation capability to remove the undesired material from the surface.
 ii. This methodology uses plasma that is less expensive than wet etching that needs a large amount of etchant chemical.
 iii. Easy to handle because it does not require the use of dangerous acids and solvents.
 iv. It produces the anisotropic and isotropic profiles while reacting with an etchant.
 v. It provides better process control, less undercut, high resolution, and cleanliness.

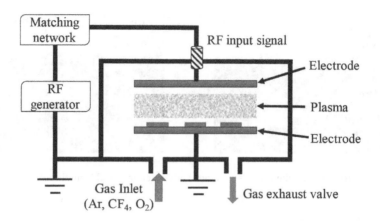

FIGURE 2.27 Dry etching process using a high RF signal.

2.2.9.2 Wet Etching

In order to perform wet etching, most of the etchant material contains oxidizing agents such as H_2O_2, Br_2, $AgNO_3/CrO_3$, HNO_3, and NaOCl. In this process, the etchant materials (such as HF, BF_6, BCL_3) are diffused on to the silicon wafer and then are reacted with each other in order to remove the unwanted materials as depicted in Figure 2.28 [67]. The chemical reaction involved in wet etching is as follows:

$$SiO_2 + 6HF \rightarrow H_2SiF_6 + 2H_2O \qquad (2.22)$$

The etching rate (R) is mainly temperature dependent that can be expressed using the Arrhenius equation by

$$R = R_0 e^{-E_A/kT} \qquad (2.23)$$

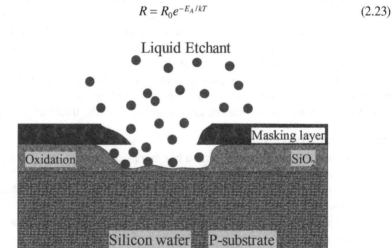

FIGURE 2.28 Wet etching process on the silicon wafer.

where R_0, E_A, k, and T are denoted as the density and diffusivity of reactant-dependent rate constant, activation energy of material, Boltzmann constant, and operating temperature, respectively.

Advantages and disadvantages

  i. The equipment used for dry etching is a simple and good selectivity for most of the materials.
 ii. It exhibits a higher etching rate and selectivity as compared to dry etching.
iii. Mostly, it is isotropic, which requires a large amount of etchant chemical to maintain the same initial etching rate and to remove unwanted material from a silicon wafer.
 iv. The cost required to perform wet etching is extremely higher as compared to dry etching.

2.3   BASIC CMOS TECHNOLOGY

In VLSI technology, the fabrication of an IC is primarily used to integrate several different components such as a resistor, transistor, diode, and capacitor within the die area based on the requirements. This can be only possible with the help of several processes such as oxide growth, epitaxial growth, masked impurity diffusion, photolithography, oxide etching, and metallization. Moreover, the transistor in its various forms, such as CMOS, BJT, and FET, etc., is the most complicated element to fabricate over a silicon wafer. The different technology approach is used based on the type of transistor built over an IC. In this section, we can familiarize with the approach to the concept of CMOS fabrication that can be accomplished based on either of the following technologies [68]

  i. N-well and P-well technology
 ii. Twin-well technology

2.3.1   N-WELL AND P-WELL CMOS PROCESS

The fabrication of CMOS transistors can be obtained by integrating NMOS and PMOS over the same silicon wafer surface by using either N-well or P-well technology. In N-well technology, the formation of CMOS transistor using N-well can be obtained, as shown in Figure 2.29 based on the procedure demonstrated below:

Step-I: This step involved the selection of a P-type silicon wafer in order to use N-well technology to form the CMOS transistor. Then, the silicon dioxide as a barrier is used to deposit on top of the silicon wafer using an oxidation process such that it gets protected from any external or internal contaminant.

Step-II: In order to perform the photolithography process, the layer of SiO_2 is coated uniformly with photoresist material with an appropriate thickness based on the technology requirements. Further, the desired pattern is produced using a stencil that is used for masking on top of the resist material.

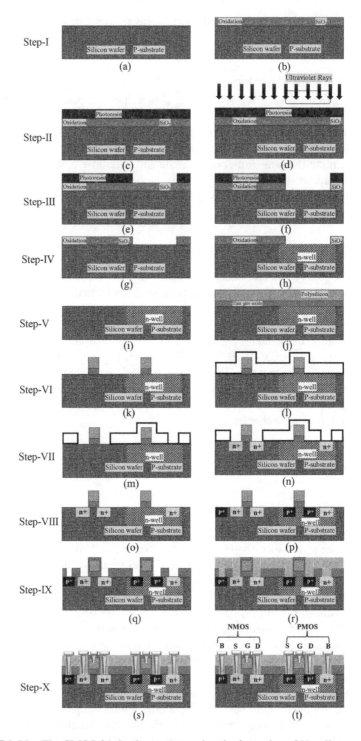

FIGURE 2.29 The CMOS fabrication process using the formation of *N*-well.

Furthermore, the ultraviolet (UV) rays are passed through photoresist material and the masked area gets polymerized.

Step-III: The unexposed region is eliminated using the chemical such as trichloroethylene. Later, the silicon wafer is dropped inside the etchant solution of hydrofluoric acid that is used to eliminate only the oxide from the selected area and the area covered by photoresist will not have any effect.

Step-IV: After the completion of the etching process, the whole photoresist mask is removed using the H_2SO_4 chemical solvent. Further, the diffusion of N-type impurities is performed in order to form N-well on top of the P-type silicon substrate.

Step-V: In the later stage of the formation of N-well, this step is further carried out by eliminating the silicon dioxide layer from the top of the silicon wafer using the hydrofluoric acid. Later, a thin layer of gate oxide is coated on a silicon wafer in order to perform the self-aligned gate process. The deposition of thin gate oxide is necessary such that misalignment of the gate should not occur that leads to the unwanted capacitance that reduces the system performance. Moreover, the polysilicon is deposited using a CVD process over thin oxide to develop the gate region such that it can sustain at high temperatures approximately 8000°C when the annealing process is performed on a silicon wafer.

Step-VI: The whole portion of the polysilicon is eliminated except the two regions where the formation of the gate region is required for the NMOS and PMOS transistors. This process is followed by an oxidation process that is used as a protective layer before executing the diffusion and metallization.

Step-VII: Further, few small regions are formed by masking the silicon wafer in order to diffuse N-type impurities on P-substrate and N-well. Moreover, the three $N+$ regions are formed using a diffusion process to create the NMOS terminal.

Step-VIII: This step is used to remove the oxide layer followed by diffusion of three $P+$ regions in order to form the PMOS terminal similar to N-type diffusion as observed in step-VII.

Step-IX: Before forming the metal terminal using the metallization process, a layer of thick oxide needs to lay out to form a protected region where no terminal needs to form. After this process, a suitable material is used to form the metal terminal in the whole wafer area in order to develop the interconnection.

Step X: In the final stage, the etching process is used to remove any excess metal present on the top of the silicon wafer and the metal terminals are formed in the gap after the etch process. Thereafter, all the terminals of NMOS and PMOS transistors are assigned with a unique name for the final product.

Similarly, P-well technology can also be understood based on the above approach, wherein the development of CMOS device followed the same procedure except for the creation of P-well diffusion instead of N-well over N-substrate. The complete layout and cross-sectional view are demonstrated in Figures 2.30 and 2.31, respectively.

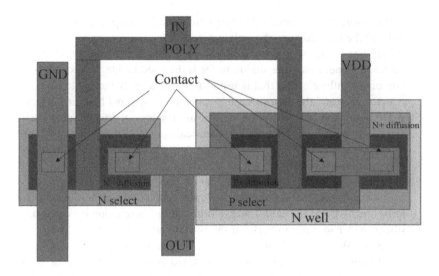

FIGURE 2.30 Layout of *P*-type CMOS.

2.3.2 TWIN-TUB PROCESS

In the twin-tub process, the separate optimization can be performed for NMOS and PMOS transistor parameters such as body effect, channel transconductance, and threshold voltage. In this technology, the primary substrate is considered either *N*-type or *P*-type substrate with a lightly doped epitaxial layer on top of it. The formation of *N*-well and *P*-well is the first step toward the beginning of the twin-tub process, wherein the formation is similar to the process as explained in Subsection 2.3.1. An appropriate dopant concentration is applied in order to produce the desired characteristics of the device. The complete dual-well or twin-tub process is demonstrated in Figure 2.32.

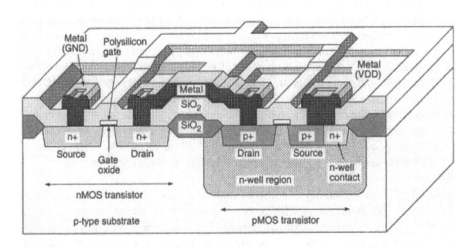

FIGURE 2.31 Cross-sectional view of CMOS device.

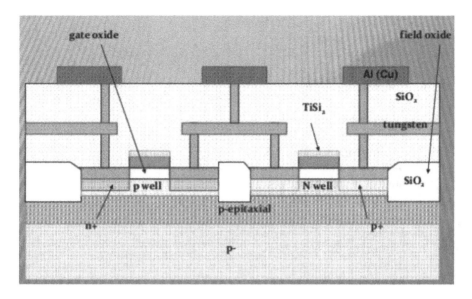

FIGURE 2.32   Cross-sectional view of *N*- and *P*-type CMOS using twin-tube process.

2.4   SUMMARY

This chapter can be summarized as follows:

- This chapter dealt with the introduction of scaling in VLSI technology followed by a detailed explanation about the fabrication process.
- A complete fabrication process was explained by demonstrating several methods based on the requirements. The techniques and methods involved in the fabrication process are purifications, wafer processing, oxidation, epitaxial, different lithography processes, deposition of poly-silicon and other dielectric materials, diffusion of atoms, ion implantation, metallization, and etching process. The whole process provides detailed information regarding the flow and process to understand the fabrication aspects.
- Further, CMOS process technology was covered and explained based on *N*-well, *P*-well, and twin-tub processes.

2.5   MULTIPLE-CHOICE QUESTIONS

1. The parameter which is not scaled to any factor is
   a. Power speed product
   b. Switching energy
   c. Channel resistance
   d. All of the mentioned

2. CMOS technology is used in developing
   a. Microprocessors
   b. Microcontrollers
   c. Digital logic circuits
   d. All of the mentioned

3. *P*-well is created on
   a. *P*-substrate
   b. *N*-substrate
   c. *p*- & *n*-substrates
   d. none of the mentioned

4. The oxidation process is carried out using
   a. Hydrogen
   b. Low purity oxygen
   c. Sulphur
   d. Nitrogen

5. The photoresist layer is formed using
   a. High sensitive polymer
   b. Light-sensitive polymer
   c. Polysilicon
   d. Silicon dioxide

6. In CMOS fabrication, the photoresist layer is exposed to
   a. Visible light
   b. Ultraviolet light
   c. Infrared light
   d. Fluorescent light

7. Few parts of the photoresist layer are removed by using
   a. Acidic solution
   b. Neutral solution
   c. Pure water
   d. Diluted water

8. Which type of CMOS circuits is good and better?
   a. *P*-well
   b. *N*-well
   c. All of the mentioned
   d. None of the mentioned

9. *N*-well is formed by
   a. Decomposition
   b. Diffusion

c. Dispersion
d. Filtering

10. _____ is sputtered on the whole wafer
    a. Silicon
    b. Calcium
    c. Potassium
    d. Aluminum

11. Heavily doped polysilicon is deposited using
    a. Chemical vapor decomposition
    b. Chemical vapor deposition
    c. Chemical deposition
    d. Dry deposition

12. Interconnection pattern is made on
    a. Polysilicon layer
    b. Silicon-di-oxide layer
    c. Metal layer
    d. Diffusion layer

13. Which is used for the interconnection?
    a. Boron
    b. Oxygen
    c. Aluminum
    d. Silicon

14. Lithography is
    a. Process used to transfer a pattern to a layer on the chip
    b. Process used to develop an oxidation layer on the chip
    c. Process used to develop a metal layer on the chip
    d. Process used to produce the chip

15. Silicon oxide is patterned on a substrate using
    a. Physical lithography
    b. Photolithography
    c. Chemical lithography
    d. Mechanical lithography

16. Positive photoresists are used more than negative photoresists because:
    a. Negative photoresists are more sensitive to light, but their photolithographic resolution is not as high as that of the positive photoresists
    b. Positive photoresists are more sensitive to light, but their photolithographic resolution is not as high as that of the negative photoresists
    c. Negative photoresists are less sensitive to light
    d. Positive photoresists are less sensitive to light

17. _____ is/are used to reduce the resistivity of polysilicon.
    a. Photoresist
    b. Etching
    c. Doping impurities
    d. None of the mentioned

18. The dopants are introduced in the active areas of silicon by:
    a. Diffusion process
    b. Ion implantation process
    c. Chemical vapor deposition
    d. Either diffusion or ion implantation process

19. To grow the polysilicon gate layer, the chemical used for CVD is:
    a. Silicon Nitride(Si_4N_3)
    b. Silane gas(SiH_4)
    c. Silicon oxide
    d. None of the mentioned

20. Gate oxide layer consists of
    a. SiO_2 layer, overlaid with a few layers of an oxynitride oxide
    b. Only SiO_2 layer
    c. SiO_2 layer with polysilicon layer
    d. SiO_2 layer and stack of epitaxial layers of polysilicon

2.6   SHORT ANSWER QUESTIONS

1. Why SiO_2 is an important component in electronics?
2. Explain the application of SiO_2 layer in IC fabrication.
3. What are salicide and silicide?
4. What do you mean by epitaxy?
5. What is the difference between pseudo homo epitaxy and heteroepitaxy?
6. What do you mean by annealing?
7. What are the PR materials?
8. What is the difference between positive and negative photoresist?
9. What is plasma?
10. Why a higher degree of anisotropy is required in VLSI fabrication?
11. How the thickness of the deposited film is measured?
12. What is electromigration?

2.7   LONG ANSWER QUESTIONS

1. What are the various types of defects in the crystal structure? Explain.
2. Describe CZ process in detail with a neat diagram. What is the Pull Rate in CZ technique? How the Pull Rate is controlled during the CZ crystal growth process?

3. What is CZ method? Explain along with its advantages and disadvantages.
4. What is Epitaxy? Discuss the MBE technique in brief. What are the advantages of MBE over VPE?
5. Describe a typical ion implanter. What are the advantages of ion implantation?
6. What is ion implantation? Explain the process with a neat diagram.
7. What do you mean by photo-resist? Explain various types of photo-resist.
8. Describe all types of PR. What are the properties of PR?
9. List and compare different types of lithography techniques.
10. Explain the ion beam lithography process.
11. List and explain all the steps of pattern transfer using the photolithography process.
12. Explain the X-ray lithography process.
13. What is X-ray lithography? Describe the advantages and problem areas associated with X-Ray lithography.
14. What do you mean by the etching process? Explain all the etching processes in brief with a neat diagram.
15. Explain the metallization and describe the problems associated with this process.
16. Why metallization is required? What advantages and applications it provides the ICs?
17. Give some solutions to get rid of the electromigration problem.
18. With neat diagram, explain the fabrication process sequence for NMOS IC technology.
19. Give the various fabrication steps of CMOS transistor using *N*-well technique with diagrams and a brief explanation.

REFERENCES

1. Bardeen, J. and Brattain, W. 1948. The transistor, a semiconductor triode. *Physics Review*, 74: 230–231.
2. Beeson, R. and Ruegg, H. 1962. New forms of all-transistor logic. *IEEE International Solid-State Circuits Conference*. Digest of Technical Papers, Philadelphia, PA, USA, pp. 10–11.
3. Riordan, M. and Hoddeson, L. 2007. Crystal fire: The invention, development and impact of the transistor. *IEEE Solid-State Circuits Society Newsletter* 12, no. 2: 24–29.
4. Kilby, J. S. 2019. Available at: https://history-computer.com/Library/US3138743.pdf (accessed December 15, 2019).
5. Wanlass, F. M. and Sah, C. T. 1963. Nanowatt logic using field-effect metal-oxide semiconductor triodes. *IEEE International Solid-State Circuits Conference*. Digest of Technical Papers, Philadelphia, PA, USA, pp. 32–33.
6. Antognetti, P., Pederson, D. O. and Man, H. D. 1981. Computer design aids for VLSI circuits. *NATO Advanced Institute*, Sigthoff & Noordhoff, Alphen aanden Rijn, The Netherlands.
7. Ruehli, A. E. 1986. *Circuit analysis. Simulation and Design*, North-Holland, New York, NY.
8. Schwarz, A. F. 1987. *Computer-Aided Design of Microelectronic Circuits and Systems*, Vols. I and II, Academic Press, New York, NY.

9. Gajski, D. and Kuhn, R. New VLSI tool, *IEEE Computer* 16, no. 12: 11–14.

10. Moore, G. E. 2006. Cramming more components onto integrated circuits, reprinted from Electronics, vol. 38, no. 8, April 19, 1965, pp. 114 ff., *IEEE Solid-State Circuits Society Newsletter* 11, no. 3: 33–35.

11. "Moore's Law". 2019. Available: http://www.intel.com/research/silicon/mooreslaw.htm (accessed December 15, 2019).

12. Weste, N. and Harris, D. 2010. *CMOS VLSI Design: A Circuits and Systems Perspective*, Pearson, p. 33.

13. Mead, C. and Conway, L. 1980. *Introduction to VLSI System*, Addison-Wesley, Reading, MA, p. 63.

14. Yaws, C. L., Dickens, L. L., Lutwack, R. and Hsu, G. 1981. Semiconductor industry silicon: Physical and thermodynamic properties. *Solid State Technology* 24, no. 1: 87–92.

15. Beadle, W. E., Tsai, J. C. C. and Plummer, R. D. 1985. *Quick Reference Manuel for Silicon Integrated Circuit Technology*, Wiley, New York, NY.

16. Crossman, L. D. and Baker, J. A. 1977. Polysilicon technology. In *Semiconductor Silicon*. Electrochemical Society, Pennington, NJ, p. 18.

17. McCormick, J. R. 1986. Polycrystalline silicon. In *Semiconductor Silicon*, Electrochemical Society, Pennington, NJ, p. 43.

18. Rea, S. N. 1981. Czochralski silicon pull rate limits. *Journal of Crystal Growth* 54, no. 2: 267–274.

19. Available: https://meroli.web.cern.ch/Lecture_silicon_floatzone_czochralski.html (accessed December 18, 2019).

20. Bonora, A. C. 1977. Silicon wafer process technology: Slicing, etching, polishing. In *Semiconductor Silicon*, Electrochemical Society, Pennington, NJ, p. 154.

21. Kern, W. 1978. Chemical etching of silicon, germanium, gallium arsenide, and gallium phosphide. *RCA Review* 39, p. 278–308.

22. Atalla, M. M. 1960. Semiconductor surfaces and films; The Si-SiO$_2$ system. In H. Gatos (Ed.), *Properties of Elemental and Compound Semiconductors* (Vol. 5), Interscience, New York, pp. 163–181.

23. Massoud, H. Z., Plummer, J. D. and Irene, E. A. 1985. Thermal oxidation of silicon in dry oxygen growth-rate enhancement in the thin regime. *Journal of Electrochemical Society* 132, no. 11: 2693–2700.

24. Kern, W. and Puotinen, D. A. 1970. Cleaning solutions based on hydrogen peroxide for use in silicon semiconductor technology. *RCA Review* 31: 187–206.

25. Ligenza, J. R. 1961. Effect of crystal orientation on the oxidation rates in High pressure steam. *The Journal of Physical Chemistry* 65, no. 11: 2011–2014.

26. Ho, V. Q. and Sugano, T. 1980. Selective anodic oxidation of silicon in oxygen plasma. *IEEE Transactions on electron devices* 27, no. 8: 1436–1443.

27. Grove, A. S., Leistiko, O. and Sah, C. T. 1964. Redistribution of acceptor and donor impurities during thermal oxidation of silicon. *Journal of Applied Physics* 35, no. 9: 2695–2701.

28. Murarka, S. P. 1975. Diffusion and segregation of ion-implanted boron in silicon in dry oxygen ambients. *Physics Review B* 12, no. 6: 2502–2519.

29. Tilli, M., Motooka, T., Airaksinen, V. M., Franssila, S., Paulasto-Kröckel, M. and Lindroos, V. 2010. *Handbook of Silicon Based MEMS Materials and Technologies*, Elsevier.

30. Pohl, U. W. 2013. *Epitaxy of Semiconductors: Introduction to Physical Principles*. Springer Science & Business Media, pp. 4–6.

31. Schreck, M., Hörmann, F., Roll, H., Lindner, J. K. N. and Stritzker, B. 2001. Diamond nucleation on iridium buffer layers and subsequent textured growth: A route for the realization of single-crystal diamond films. *Applied Physics Letters* 78, no. 2: 192–194.

32. Ban, V. S. 1977. Mass spectrometric studies of chemical reactions and transport phenomena in silicon epitaxy. In *Proceeding of 6th International CVD Conference of Atlanta*, p. 66.
33. Morgan, D. V. and Board, K. 1991. *An Introduction to Semiconductor Microtechnology* (2nd ed.), John Wiley & Sons, Chichester, West Sussex, England, p. 23.
34. Cho, A. Y. and Arthur, J. R. 1975. Molecular beam epitaxy. *Progress in Solid State Chemistry*, 10: 157–192.
35. Bean, J. C. 1981. Growth of doped silicon layers by molecular beam epitaxy. In F.F.Y. Wang (Ed.), *Impurity Doping Processes in Silicon*, Amsterdam, The Netherlands.
36. Available: https://capricorn.bc.edu/wp/zeljkoviclab/research/molecular-beam-epitaxy-mbe/ (accessed December 18, 2019).
37. Colinge, J. P. 2003. Multiplate gate silicon on insulator MOS transistors. In *Microelectronics Technology and Devices, SBMICRO 2003: Proceedings of the Eighteenth International Symposium. The Electrochemical Society*, pp. 2–17.
38. Meggs, P. B. 1998. *A History of Graphic Design*, John Wiley & Sons, New York, pp. 146–148.
39. Dill, F. H., Hornberger, W. P., Hauge, P. S. and Shaw, J. M. 1975. Characterization of positive photoresist. *IEEE Transactions on Electron Devices* 22, no. 7: 445–452.
40. Thompson, L. F. 1983. *An Introduction to Lithography, Introduction to Microlithography*, American Chemical Society, Washington, DC.
41. Pala, N. and Karabiyik, M. 2016. Electron beam lithography (EBL). In Bhushan B. (Ed.), *Encyclopaedia of Nanotechnology*, Springer, Dordrecht.
42. Parker, N. W., Brodie, A. D. and McCoy, J. H. 2000. High-throughput NGL electron-beam direct-write lithography system. *Proceeding of SPIE. Emerging Lithographic Technologies IV* 3997: 713–720.
43. Spears, D. L. and Smith, H. I. 1972. High-resolution pattern replication using soft x-rays. *Electronics Letters* 8, no. 4: 102–104.
44. Peckerar, M. C. and Maldonado, J. R. 1993. X-ray lithography—An overview. *In Proceedings of the IEEE* 81, no. 9: 1249–1274.
45. Kern, W. and Ban, V.S. 1978. Chemical vapor deposition of inorganic thin films. In *Thin Film Process*, Academic Press, New York, pp. 257–331.
46. Kern, W. and Schnable, G. L. 1979. Low pressure chemical vapour deposition for very large scale integration processing—A review. *IEEE Transactions on Electron Devices* 26, no. no. 4: 647–657.
47. Hitchman, M. L. and Ahmed, W. 1984. Some recent trends in the preparation of thin layers by low pressure chemical vapour deposition. *Vacuum* 34, no. 10–11: 979–986.
48. Lin, M. B. 2011. *Introduction to VLSI Systems—A Logic Circuit, and System Perspective*, CRC Press, Taylor & Francis Group, pp. 151–157.
49. Wang, J. T. 2011. CVD and its related theories in inorganic synthesis and materials preparations. *Modern Inorganic Synthetic Chemistry*. 151–171.
50. Morgan, D. V. and Board, K. 1991. *An Introduction to Semiconductor Microtechnology* (2nd edition), John Wiley & Sons, Chichester, West Sussex, England, p. 27.
51. Appels, J. A., Kooi, E., Paffen, M. M., Schatorji, J. J. H. and Verkuylen, W. H. C. G. 1970. Local oxidation of silicon and its application in semiconductor device technology. *Philips Research Report*, 25, 118–132.
52. Makino, T. 1983. Composition and structure control by source gas ratio in LPCVD SiNx. *Journal of Electrochemical Society* 130, no. 2: 450–455.
53. Pan, P. and Berry, W. 1985. The composition and physical properties of LPCVD silicon nitride deposited with different NH_3/SiH_2Cl_2 gas ration. *Journal of Electrochemical Society* 132, no. 12: 3001–3005.

54. Philibert, J. 2005. One and a half century of diffusion: Fick, Einstein, before and beyond. Archived 2013-12-13 at *The Wayback Machine Diffusion Fundamentals* 2: 1.1–1.10.

55. Fick, A. 1855. Ueber diffusion. *Annalen der Physik (in German)* 94, no. 1: 59–86.

56. Swaminathan, B., Saraswat, K. C. and Dutton, R. W. 1982. Diffusion of arsenic in polycrystalline silicon. *Applied Physics Letters* 40, no. 9: 795–798.

57. Kamins, T. I., Manoliu, J. and Tucker, R. N. 1972. Diffusion of impurities in polycrystalline silicon. *Journal of Applied Physics* 43, no. 1: 83–91.

58. Horiuchi, S. and Blanchard, R. 1975. Boron diffusion in polycrystalline silicon layers. *Solid-State Electronics* 18, no. 6: 529–532.

59. Würker, W., Roy, K. and Hesse, J. 1974. Diffusion and solid solubility of chromium in silicon. *Materials Research Bulletin* 9, no. 7: 971–977.

60. Bailey, R. F. and Mills, T. G. 1969. Diffusion parameters of platinum in silicon. In R. R. Habarecht and E. L. Kern (Eds.), *Semiconductor Silicon*. Electrochemical Society, New York, p. 481.

61. Sze, S. M. 2003. *VLSI Technology* (2nd edition), McGraw Hill Education, p. 329

62. Rose, P. H. 2006. Concepts and designs of ion implantation equipment for semiconductor processing. *Review of Scientific Instruments* 77, no. 11: 111101.1–111101.12.

63. Xiao, Hong. 2001. *Introduction to Semiconductor Manufacturing Technology*, Pearson Prentice Hall, p. 453.

64. Park, S., Lee, Y., Kim, B. S. and Kim, I. S. 2012. Thermal insulation property of cu-metallized nanofibers. *Advances in Polymer Technology* 31, no. 1: 1–6.

65. Sze, S. M. 2003. *VLSI Technology* (2nd edition), McGraw Hill Education, p. 413.

66. Forgotson, N., Khemka, V. and Hopwood, J. 1996. Inductively coupled plasma for polymer etching of 200 mm wafers. *Journal of Vacuum Science & Technology B: Microelectronics and Nanometer Structures Processing, Measurement, and Phenomena* 14, no. 2: 732–737.

67. Zhang, X. and Hoshino, K. 2019. Fundamentals of nano/microfabrication and scale effect. *Molecular Sensors and Nanodevices*, 43–111.

68. Lin, M. B. 2011. *Introduction to VLSI Systems—A Logic Circuit, and System Perspective*, CRC Press, Taylor & Francis Group, p. 45.

3 MOSFET Modeling

3.1 INTRODUCTION TO MOS TRANSISTOR

Complete knowledge about the metal oxide semiconductor field-effect transistor (MOSFET) is the primary task toward establishing strong steps in the field of very-large-scale integration (VLSI) technology [1]. In integrated circuits (ICs), the metal oxide semiconductor (MOS) transistor is fabricated on a single chip due to its small die area and acts as the core element of an IC. It is widely used for switching and amplification of signal that consists of four terminals such as *source* (*S*), *gate* (*G*), *drain* (*D*), and *body* (*B*), as depicted in Figure 3.1 [2,3]. Moreover, the source and drain are formed on either side on top of the silicon substrate by using heavily doped atoms, whereas the gate terminal is created using a metal or heavily doped polysilicon on top of the thin oxide layer [1]. The structure of MOS is symmetrical and due to symmetrical structure, the source and drain terminal cannot be distinguished until the terminals are biased. The gate terminal acts as a controlling electrode by varying the channel width that primarily depends on the applied voltage. The applied voltage at the gate terminal is basically responsible for the flow of charge (electrons and holes) carriers from source to drain under gate control. The gate terminal is placed on top of the thin metal oxide layers in order to provide insulation from the channel.

In order to understand the function of MOS transistor, the gate terminal is triggered with a biased supply voltage that generates an electric field resulting in the flow of charge carriers from source to drain. However, the electric field generated vertically is controlled by the gate voltage and lateral field via source to drain, the device is termed as MOSFET. Moreover, it is also termed as insulated gate field-effect transistor (IGFET) because of the gate terminal placed on top of the insulating thin oxide layer [2, 3].

Based on the mode of operation, MOSFET can either work in *depletion mode* or in *enhancement mode*. In addition to that, the formation of a channel between the source and drain can act as an *N*-type or *P*-type MOSFET depending on the carriers doped in the channel region. If the channel region is formed with *N*-type material having the flow of electron from source (heavily doped with *N*+) to drain (*P*-type substrate) region, it acts as *N*-type MOSFET. Similarly, if the channel region is formed with *P*-type material having a flow of holes from source (heavily doped with *P*+) to drain region (*N*-type substrate), it acts as *P*-type MOSFET [3, 4]. The mode of operation in which the different type of MOS transistor operates as ON and OFF based on gate voltage is summarized in Table 3.1.

> **Depletion-mode MOSFET:** Since the gate terminal of MOSFET is unbiased, the conductivity between the channels of MOSFET remains at maximum, whereas the conductivity of channel reduces under positive- or negative-biased conditions. This phenomenon is due to the formation of layer between the source and drain during the fabrication using ion implantation that

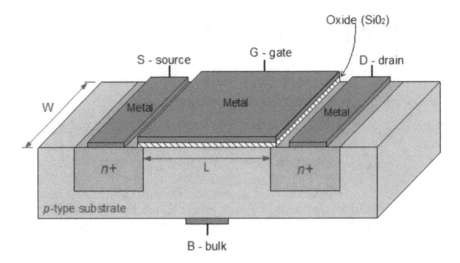

FIGURE 3.1 The most common structure of an *N*-channel MOS transistor in three dimensions.

keeps it always in ON condition until the gate terminal is applied with bias voltage in order to turn it OFF. The symbolic representation of *N*-channel and *P*-channel depletion-mode MOSFET is depicted in Figure 3.2. This phenomenon can be obtained in the depletion-mode MOSFET.

Enhancement-mode MOSFET: However, in the case of enhancement mode of MOSFET, the conductivity across the channel can only be observed during the applied voltage at the gate terminal; otherwise no conduction occurs under zero-biased voltage. The symbolic representations of *N*-channel/*P*-channel MOSFET under depletion and enhancement mode operation are depicted in Figure 3.2.

3.1.1 CHARACTERISTICS OF MOS TRANSISTOR

The characteristics of MOS transistors are presented for the NMOS (*N*-channel) device that is also valid for the PMOS (*P*-channel) device. In a MOS device, the terminals are assigned with unique voltage notations such as V_g, V_s, V_d, and V_b as

TABLE 3.1

Types of MOSFET Having *N*-type and *P*-type Channel

Devices	Operating State	Applied Voltage as Gate terminal for	
		N-channel	*P*-channel
Depletion type	ON	−Vg Turns OFF	+Vg Turn OFF
Enhancement type	OFF	+Vg Turns ON	−Vg Turn ON

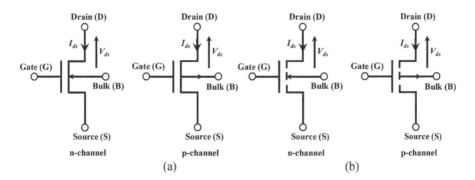

FIGURE 3.2 Symbolic representation of *N*-channel/*P*-channel MOSFET under (a) depletion condition and (b) enhancement condition.

the voltage at the gate, source, drain, and body terminals, respectively. Similarly,
the drain source, gate source, and bulk source are denoted as V_{ds} (= $V_d - V_s$), V_{gs}
(= $V_g - V_s$), and V_{bs} (= $V_b - V_s$), respectively [3]. The threshold voltage is the minimum voltage used to turn ON the transistor that depends on the concentration of
the majority charge carrier in the semiconductor substrate. In order to turn on the
N-channel MOSFET, the gate terminal is applied with positive voltage to create
the inversion charge. Since the gate voltage (with respect to the source) is greater
than the threshold voltage, the current flows through drain to source. Further,
the formation of the electron inversion layer takes place once the drain-to-source
voltage is applied, as shown in Figure 3.3(a). Once the electron inversion layer is
formed, the electron flows from source-to-drain terminal; hence, the transistor
remains in ON condition [5]. Similarly, Figure 3.3(b) demonstrates that the flow of
current from drain to source is zero until the voltage at the gate terminal remains
less than the threshold voltage that does not form the electron inversion layer.
Hence, the transistor remains OFF by introducing an extremely high resistance
between drain and source.

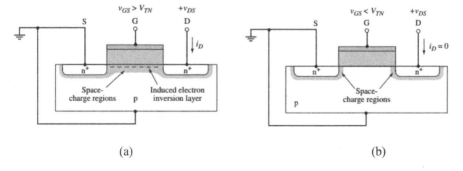

FIGURE 3.3 The impact of gate voltage for *N*-channel MOSFET under (a) $V_g > V_{th}$ condition
and (b) $V_g < V_{th}$ condition.

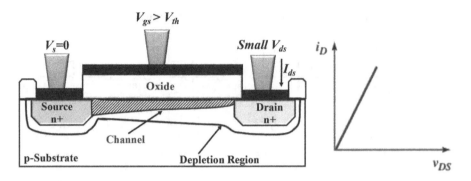

FIGURE 3.4   The *N*-channel MOSFET under threshold voltage greater than gate voltage.

The characteristics of an *N*-channel MOSFET can be defined based on the regions of operation such as the (1) linear region (2) saturation region, and (3) cut-off region.

1. **Linear region:** In this region, the flow of drain current (I_{ds}) linearly increases with an increase in the drain-to-source (V_{ds}) voltage under applied gate voltage (V_{gs}) (with respect to the source) greater than the threshold voltage (V_{th}) as depicted in Figure 3.4 [5]. The drain current expression can be given as a function of V_{ds} and V_{gs}

$$I_{ds} = \mu_n C_{oxi}\left(\frac{W}{L}\right)\left(V_{gs} - V_{th} - 0.5V_{ds}\right)V_{ds} \qquad (3.1)$$

$$I_{ds} = k'_n\left(\frac{W}{L}\right)\left(V_{gs} - V_{th} - 0.5V_{ds}\right)V_{ds} = k_n\left(V_{gs} - V_{th} - 0.5V_{ds}\right)V_{ds} \qquad (3.2)$$

where k_n is known as the *gain factor* that can be defined as the product of the *process transconductance* (k'_n) parameter and the (*W/L*) ratio of an *N*-channel MOSFET device. It can be expressed as

$$k_n = k'_n\left(\frac{W}{L}\right) = \mu_n C_{oxi}\left(\frac{W}{L}\right) = \frac{\mu_n \varepsilon_{ox}}{t_{ox}}\left(\frac{W}{L}\right) \qquad (3.3)$$

where the parameters μ_n and C_{oxi} are denoted as the mobility of charge carrier and per unit area gate oxide capacitance, respectively.

    Additionally, further increase in the applied drain-to-source voltage reduces the electron inversion layer near the drain terminal due to a decrease in the voltage across the oxide layer close to the terminal, as depicted in Figure 3.5 [5].

2. **Saturation region:** In the case, when gate voltage is greater than the threshold voltage, the drain current starts increasing as soon as the drain-to-source voltage is applied due to the formation of an electron inversion layer. But after a certain value of V_{ds}, the drain current virtually remains

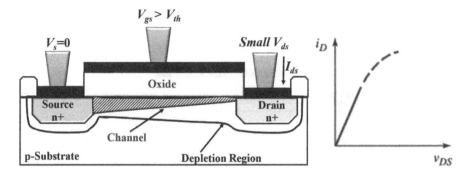

FIGURE 3.5 The *N*-channel MOSFET under threshold voltage greater than the applied gate voltage with large drain voltage.

constant even after increasing the drain-to-source voltage and that point is called saturation point or saturation voltage V_{ds}(sat), as observed in Figure 3.6 [5]. Such a phenomenon is due to an increase in the depletion region near the drain terminal and no channel is inverted. Hence, the channel is said to be under the pinched-off condition that provides constant drain current that is independent of V_{ds}. Moreover, the sudden rise in the drain current can be observed with a further increase in the value of V_{ds} until the *PN* junction between the drain and the substrate breaks down due

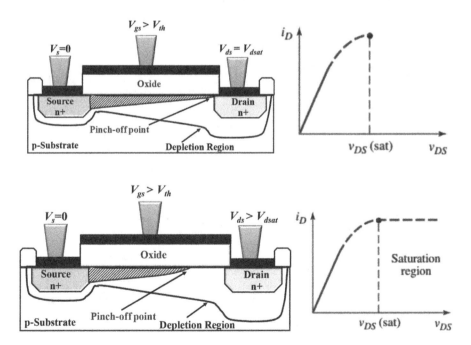

FIGURE 3.6 The behavior of *N*-channel MOSFET under saturation region ($V_{gs} > V_{th}$).

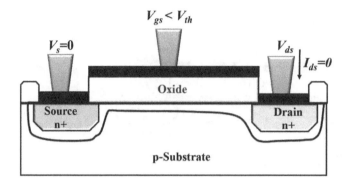

FIGURE 3.7   The behavior of *N*-channel transistor under the cut-off region ($V_{gs} < V_{th}$).

to the high electric field at the drain terminal. This phenomenon is termed as a *break down region*, as observed in Figure 3.7.

$$I_{ds} = \frac{1}{2}k'_n\left(\frac{W}{L}\right)\left(V_{gs} - V_{th}\right)^2 = \frac{k_n}{2}\left(V_{gs} - V_{th}\right)^2 \tag{3.4}$$

3. **Cut-off region:** Figure 3.8 demonstrates the behavior of *N*-channel MOS transistors under gate to source voltage less than the threshold voltage that results in no channel formation between the source and the drain [5].

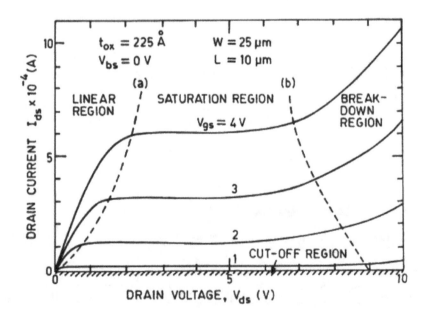

FIGURE 3.8   Different regions of operations of an *N*-channel MOS transistor [5, 6].

Therefore, the current flow is zero from drain to source. The drain current under cut-off region can be obtained using the below equation

$$I_{ds} = k'_n \left(\frac{W}{L}\right)(\eta - 1)V_t^2 exp\left[\frac{V_{gs} - V_{th}}{\eta V}\right] \tag{3.5}$$

3.1.2 Hot Carrier Effects

The performance of an electronic device degrades because of the hot carrier effect that primarily occurs for the rapid dimensional scaling of technology compared to the supply voltage. The demand for high operating frequency and miniaturization of technology can be obtained by reducing the channel length between the source and the drain while keeping the constant supply voltage. As a result of reduced channel length, the electrons and holes experience a high electric field in the lateral and vertical fields as depicted in Figure 3.9 [6,7]. The carriers within the channel acquired higher kinetic energy that penetrates into the gate oxide, which changes the charge distribution permanently near the interface [8]. Therefore, hot carrier injection is primarily responsible for the degradation of device performance and I–V characteristics of semiconductors.

3.1.3 Parasitics of MOSFET

In order to evaluate the performance of the device, the parasitic associated with the device needs to consider carefully in order to demonstrate the accurate analysis. The source/drain junction of MOSFET exhibits resistive and capacitive components that limit the performance and their impact should be minimized while designing the device.

1. **Source and drain resistance (R_{source} and R_{drain}):** The impact of the resistive component of source and drain is assumed to be neglected compared to channel resistance ($R_{channel}$) while obtaining the first-order drain

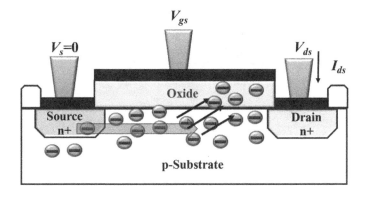

FIGURE 3.9 A cross-sectional view of the hot carrier effect.

current as described through equations (3.1) to (3.5). The expression (3.7) is obtained by differentiating (3.1), which demonstrates the channel resistance is directly proportional to the channel length (L) and increases with the longer channel.

$$R_{channel} = \frac{1}{g_{ds}} = \left(\frac{\partial I_{ds}}{\partial V_{ds}}\bigg|_{V_{gs}} \right)^{-1} \Omega \qquad (3.6)$$

$$R_{channel} = \frac{\left(V_{gs} - V_{th} - 0.5V_{ds}\right)}{\mu C_{oxi}}\left(\frac{L}{W}\right)\Omega \qquad (3.7)$$

The above expression demonstrates that the reduction of channel length can provide the behavioral change due to source and drain resistance and cannot be further neglected for shorter channel length. Moreover, the source (drain) resistance can be expressed as the sum of three factors, as depicted in Figure 3.10 [9]. These factors are R_{sheet} near the heavily doped source diffusion region, $R_{spreading}$ at the end of the source (drain) terminal, and $R_{contact}$ between the metal and source (drain) diffusion region [10]. It can be expressed as

$$R_{source} = R_{sheet} + R_{contact} + R_{spreading} \ \Omega \qquad (3.8)$$

where $R_{sheet} = \frac{\rho_s S}{W}$, $R_{contact} = \frac{\sqrt{\rho_c \rho_s}}{W} coth\left(l_c\sqrt{\frac{\rho_s}{\rho_c}}\right)$, and $R_{spreading} = \left(\frac{2\rho_s x_j}{\pi W}\right)\ln\left(H\frac{x_j}{t_{ac}}\right)$ are the sheet, contact, and spreading resistances of MOSFET, respectively [11, 12]. The parameters X_j, ρ_c, S, t_{ac}, l_c, and ρ_s are denoted as the junction

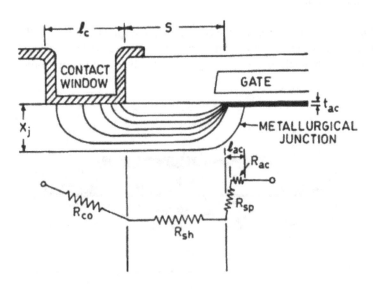

FIGURE 3.10   The equivalent resistive components of the device [9].

depth, contact resistivity between metal and source (drain) region, the distance between the contact via and the channel region, thickness of surface accumulation layer, contact length, and sheet resistance per square, respectively.

2. **Impact of resistance on device transconductance:** The expression for transconductance can be obtained using Kirchhoff's law in Figure 3.11 by considering the R_{source} and R_{drain} independent of biased voltage [13, 14]. The intrinsic drain and gate voltage are denoted as V'_{ds} and V'_{gs}, while the voltage applied is referred as V_{ds} and V_{gs} for an external terminal [15, 16].

$$V'_{gs} = V_{gs} - I_{ds}R_{source} \tag{3.9}$$

$$V'_{ds} = V_{ds} - I_{ds}\left(R_{source} + R_{drain}\right) \tag{3.10}$$

Based on the above expression, if the derivative of V_{ds} is zero at constant V_{ds}, then drain current can be written as

$$dV'_{gs} = dV_{gs} - dI_{ds}R_{source} \tag{3.11}$$

$$dV'_{ds} = -dI_{ds}\left(R_{source} + R_{drain}\right) \tag{3.12}$$

$$dI_{ds} = \left.\frac{\partial I_{ds}}{\partial V'_{gs}}\right|_{V'_{ds}}.dV'_{gs} + \left.\frac{\partial I_{ds}}{\partial V'_{ds}}\right|_{V'_{gs}} dV'_{ds} \tag{3.13}$$

$$dI_{ds} = g'_i dV'_{gs} + g'_d dV'_{ds} \tag{3.14}$$

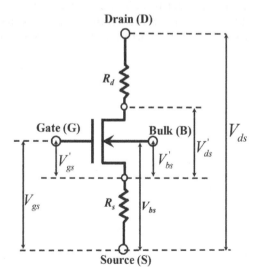

FIGURE 3.11 The source and drain resistance are mentioned as R_s and R_d, respectively in the MOSFET.

where the intrinsic and drain transconductance are denoted as g_i' and g_d', respectively. Now substituting equations (3.11) and (3.12) into (3.14), we obtain

$$g_i = \frac{g_i'}{1 + \left(R_{source} + R_{drain} \right) g_{ds}' + R_{source} g_i'} \tag{3.15}$$

Assuming that $R_{source} = R_{drain} = 0$, the presence of source (drain) resistance reduces g_i.

$$g_i = \frac{g_i'}{1 + R_{source} g_i'} \tag{3.16}$$

The expression (3.16) demonstrated that g_i is degraded in a saturation region by a factor of $\left(1 + R_{source} g_i' \right)$. It can also be observed from equation (3.15) that source and drain resistance have a higher impact on transconductance as compared to the saturation region for the nonzero value of g_{ds}' in the linear region.

3.1.4 MOSFET Circuit Models

Figure 3.12 demonstrates an equivalent DC model of MOSFET that depends on the drain current, I_{ds}, and its detailed analysis based on different models is discussed in Section 3.3. The source and drain terminals are denoted by S and D and their internal nodes are defined by S' and D', respectively, for an equivalent electrical model of MOSFET as depicted in Figure 3.12(a). In the case of DC analysis, the gate (denoted as G) terminal is assumed to be an open circuit due to separation of gate terminal from source and drain via an insulator material (e.g., gate oxide). Moreover, the DC current from base to source, base to drain, and drain to source is denoted by I_{bs}, I_{bd}, and I_{ds}, respectively, which is the nonlinear function of the terminal voltages [17]. Apart from that, Figure 3.12(b) represents the

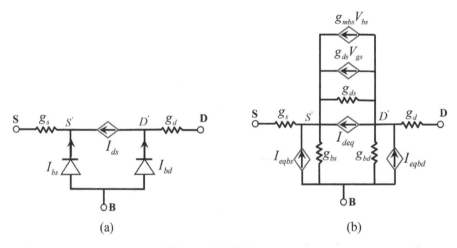

FIGURE 3.12   An (a) equivalent and (b) small-signal circuit model of MOSFET.

linear equivalent circuit having small-signal MOSFET intrinsic transconductance related to the large-signal model expressed as

$$g_i = \left.\frac{\partial I_{ds}}{\partial V_{gs}}\right|_{op}, g_{ds} = \left.\frac{\partial I_{ds}}{\partial V_{ds}}\right|_{op}, g_{ids} = \left.\frac{\partial I_{ds}}{\partial V_{gs}}\right|_{op} \tag{3.17}$$

where op denotes the operating point-biased value used to define the independent variables such as V_{gs}, V_{ds}, and V_{bs} and their associated quantitative values are assumed at op point.

3.2 MOS CAPACITOR

In this section, the impact of biased and nonbiased voltages is explored for MOS capacitor. Further, the relationship between the capacitance-voltage curve is characterized that will be used later while designing the MOS transistor.

3.2.1 MOS Capacitor With Zero and Nonzero Bias

The MOS capacitor can be formed using a sandwich structure consisting of metal, oxide, and semiconductor as depicted in Figure 3.13. The thin oxide layer is used between the metal (or polysilicon) at the top and the semiconductor substrate (either N-type or P-type) at the bottom. In order to form an ohmic contact, a second metal layer is used below the substrate that acts as the body and it will be grounded when the top metal layer is biased with gate voltage [4]. This structure can form a parallel plate capacitor if the substrate achieved enough conduction in order to provide current displacement. In this configuration, the gate terminal acts as one electrode and N-type (or P-type) substrate acts as another electrode that is separated from the

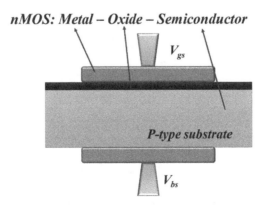

FIGURE 3.13   The structure of MOSFET.

dielectric material of SiO_2; this phenomenon is called MOS capacitor [5]. As we know from the basic concept of a parallel plate capacitor that the per unit area oxide capacitance (C_{oxi}) can also be defined as

$$C_{oxi} = \frac{\varepsilon_0 \varepsilon_{oxi}}{t_{oxi}} \tag{3.18}$$

where ε_{oxi} and t_{oxi} are the dielectric constant and thickness of oxide, respectively, in MOS technology.

Moreover, the mode of operation of a MOS capacitor can be explained using flat-band diagram, depletion, inversion, and accumulation.

1. **Flatband diagram:** The energy band diagram of aluminum metal (or poly-silicon) silicon dioxide semiconductor is depicted in Figure 3.14. During zero-biased voltage, the charge present in the semiconductor is zero and hence the semiconductor band can be observed as flat and it is termed as flatband. In order to obtain the flatband diagram, the voltage (V_{FB}) must be applied as illustrated in Figure 3.14. The characteristics of metal can be defined by its work function ϕ_M in eV that defines the energy difference between the energy of vacuum and Fermi energy of metal, $E_{F,M}$ ($\phi_M = E_{vacuum} - E_{F,M}$). The work function primarily depends on the distribution of charge or atom involved under negligible contamination [18, 19]. Further, the energy difference can be specified in terms of electron affinity χ between the vacuum and the conduction band at the surface of the semi-conductor and insulator.

2. **Accumulation:** When the gate terminal is biased with a gate voltage that is less than the flatband voltage, V_{FB}, then the negative charge appears on the gate terminal, and because of this, the positive charge appears on

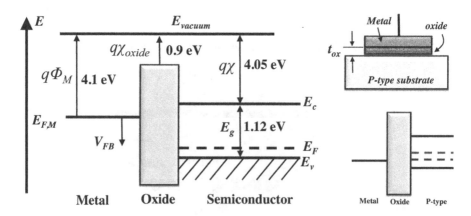

FIGURE 3.14 The flatband diagram of metal oxide semiconductor.

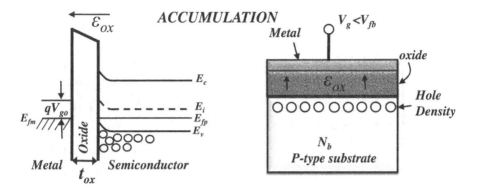

FIGURE 3.15 The accumulation of holes at negative voltage under applied voltage for *P*-type MOS capacitor.

the silicon surface. This additional positive charge gives rise to the hole concentration in the *P*-type substrate as depicted in Figure 3.15. As these excess holes accumulate at the surface, $(E_i - E_F)$ must increase, which is responsible for the bending of band upward as depicted in Figure 3.15, and the phenomenon remains in the accumulation condition.

3. **Depletion:** In this case, the gate terminal is biased with a positive gate voltage greater than the flatband voltage that induces the negative charge on the silicon surface as illustrated in Figure 3.16. The presence of negative charge on the silicon surface is responsible for the depletion of positive charge, and it is termed as depletion condition.

4. **Inversion:** This condition can be observed when the positive gate voltage is further increased, which gives rise to a more negatively charged carrier on the silicon surface. That results in further bending the band downward

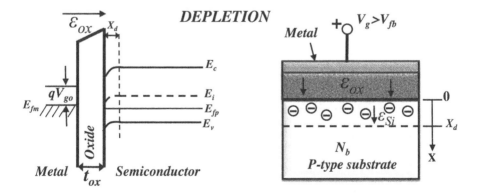

FIGURE 3.16 The depletion effect under positive voltage for *P*-type MOS capacitor.

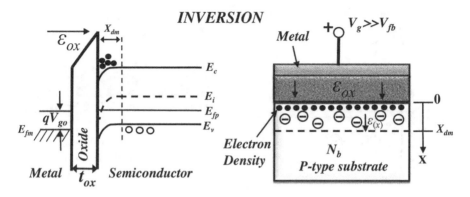

FIGURE 3.17   The inversion effect due to large applied voltage.

as depicted in Figure 3.17 [20]. Due to higher positive gate voltage greater than the flatband voltage, the *P*-type surface behaves like *N*-type surface with a rise in the electron density, this phenomenon of inversion is due to the applied voltage and hence it is referred to as inversion condition.

3.2.2 Capacitance-Voltage Curves

The capacitance-voltage (*C–V*) curve of MOS capacitor can be obtained for low frequency and high frequency as depicted in Figure 3.18 [21–23]. The charge present in the semiconductor is differentiated with respect to the potential across the semiconductor in order to obtain the low frequency or quasi-static capacitance. The calculation of capacitance must be obtained while maintaining the equilibrium condition at all times [22]. Similarly, the high-frequency capacitance can be obtained by triggering a small AC voltage in addition to the DC gate voltage.

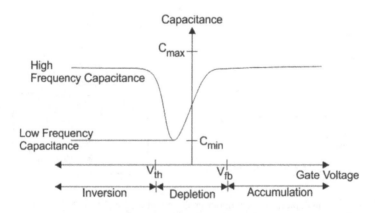

FIGURE 3.18   The capacitance-voltage curve of MOS capacitor under three region.

3.2.3 Anomalous Capacitance-Voltage Curves

As we know that the gate terminal is formed using ion implantation in submicron technology, with a very thin dimension of gate oxide (in the order of 100 Angstrom and lower), it may not be degenerately doped depending upon the process conditions [24]. As it is non-degenerately doped, it cannot be further treated as an equipotential area. The MOS capacitance can be measured using the expression

$$C_g = \left(\frac{1}{C_{oxi}} + \frac{1}{C_s} \right)^{-1} \qquad (3.19)$$

where C_s and C_g are the space region capacitance and the total MOS structure capacitance, respectively. Therefore, the MOS capacitance needs to measure by considering the impact of capacitance that occurs due to the polysilicon gate. It can be obtained as

$$C_g = \left(\frac{1}{C_{oxi}} + \frac{1}{C_s} + \frac{1}{C_{polys}} \right)^{-1} \qquad (3.20)$$

3.3 MOSFET DC AND DYNAMIC MODELS

This section discusses the static and dynamic behaviors under different operating regions while simulating the circuit based on the MOSFET models.

3.3.1 Pao-Sah Model

In 1966, Pao-Sah developed a surface potential model to calculate the drain current characteristics of MOS transistor, and the Pao-Sah model functions on the basis of the given expression [25]

$$I_{ds} = \mu_s \frac{W}{L} C_{oxi} \gamma \int_{V_{sb}}^{V_{sb}+V_{ds}} \int_{\phi_f}^{\phi_s} \frac{e^{(\phi - 2\phi_f - V_{cb})/V_t}}{F(\phi, \phi_f, V_{cb})} d\phi dV_{cb} \qquad (3.21)$$

where ϕ_s, γ, V_{fb}, μ_s, V_{cb}, V_{sb}, V_{ds}, and V_t are the surface potential, body factor, the flat band voltage, channel mobility, channel potential, source potential, drain potential, thermal voltage, and bulk-fermi potential, respectively. The double integral equation in equation (3.21), which can only be computed numerically, not only includes the drift but also has a diffusion component of the drain current. Therefore, the Pao-Sah model is valid for all the regions of the device operation. However, the computation time required for this model is large due to double integral operation and complexity is high that is not suitable for circuit simulators.

3.3.2 Charge Sheet Model

In 1978, the charge sheet model was derived first by R. Brews while assuming the inversion layer as a simple charge sheet [26]. The analysis was carried out by considering the zero thickness of the inversion layer such that the potential drop is zero. In addition to that, it is also assumed that "the depletion region under the gate is practically free of mobile carriers so that the depletion approximation is valid" [26, 27]. The drain current (I_{ds}) is the sum of the current due to drift components and diffusion components that depend on the strong and weak inversion.

$$I_{ds} = I_{ds1} + I_{ds2} \tag{3.22}$$

The below expression is valid only if the drift current is present,

$$I_{ds1} = \mu_s \frac{W}{L} C_{oxi} \left[\left(V_{gb} - V_{fb} \right) \left(\phi_{sL} - \phi_{s0} \right) - \frac{1}{2} \left(\phi_{sL}^2 - \phi_{s0}^2 \right) - \frac{2}{3} \gamma \left\{ \phi_{sL}^{3/2} - \phi_{s0}^{3/2} \right\} \right] \tag{3.23}$$

Similarly, the below expression is valid if the diffusion component is assumed to be presented

$$I_{ds2} = \mu_s \frac{W}{L} C_{oxi} V_t \left[\gamma \left(\phi_{sL}^{1/2} - \phi_{s0}^{1/2} \right) + \left(\phi_{sL} - \phi_{s0} \right) \right] \tag{3.24}$$

where ϕ_{s0} and ϕ_{sL} are the values of the source and drain terminal of the channel, respectively. In the case of strong inversion ($I_{ds} \approx I_{ds1}$), the current present in the device is due to drift, and for the weak inversion ($I_{ds} \approx I_{ds2}$), the current is be due to diffusion.

3.3.3 Piece-Wise Model for Enhancement Devices

The drain current in all the regions such as saturation, linear, and cut-off can be given based on the first-order MOSFET model of enhancement type [28].

$$I_{ds} = \begin{cases} 0, \; V_{gs} \leq V_{th} \; (Cutoff \; region) \\ \beta \left(V_{gs} - V_{th} - 0.5 V_{ds} \right) V_{ds} \; V_{gs} > V_{th}, V_{ds} \leq V_{dsat} \; (Linear \; region) \\ \frac{\beta}{2} \left(V_{gs} - V_{th} \right)^2 \left(1 + \lambda V_{ds} \right) \; V_{gs} > V_{th}, V_{ds} > V_{dsat} \; (Saturation \; region) \end{cases} \tag{3.25}$$

The above expression is derived and valid based on the following assumptions:

a. In NMOS transistor, the current due to hole charges can be neglected and the gradual channel approximation (GCA) is considered to be valid.
b. The flow of current in a MOS transistor is considered only along the length of the channel, i.e., in the y-direction, however, the recombination and generation are neglected.
c. The carrier mobility, μ_s, is kept constant along the channel length in the inversion layer and the current flow is only considered due to drift while neglecting the diffusion current.
d. The bulk charge Q_b is constant at any point along the channel length.

3.3.4 SMALL GEOMETRY MODEL

In 1988, S. Veeraraghavan and Fossum proposed a physical model that is applicable for devices and circuits. In most of the analysis, the length and width of the device are considered large enough to have a negligible impact of edge effect. Although the device exhibits several characteristics, the physical dimensions are reduced [29]:

  a. For a long-channel device, the drain current increases with an increase in the drain voltage. For short-channel lengths of the device, the higher slope is observed and exhibits a softer breakdown that generally do not occur in the longer-channel devices as depicted in Figure 3.19 [30]. This phenomenon was observed as the first short-channel effect.
  b. For longer-channel devices, the threshold voltage is geometrically dependent, while it is drain-current dependent in case of short-channel devices due to drain-induced barrier lowering (DIBL) effect.
  c. When the device length is further reduced to a narrow dimension, the gate terminal no longer has control over the drain current and the device cannot be turned off.

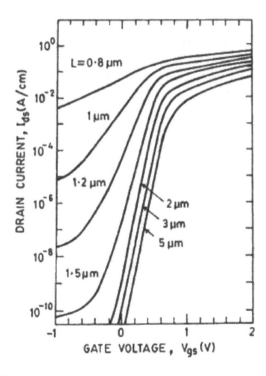

FIGURE 3.19 The subthreshold characteristics of a MOSFET under various channel lengths [30].

In order to model the DC characteristics of MOS transistor having small geometry, the following effects are required to be considered:

1. The reduction in surface mobility needs to consider, which occurs due to vertical and lateral fields.
2. Carrier velocity saturation and channel length modulation.
3. The depletion of charge sharing by the source and drain terminal.
4. Hot carrier effect and source-drain series resistance.

3.3.5   INTRINSIC CHARGES AND CAPACITANCE

The current generated in the MOS transistor flows from source to drain only due to the mobile charge carrier (holes and electrons are responsible for the flow of current in PMOS and NMOS, respectively) under steady-state condition. The additional current at the device is termed as charging current that is associated with the stored charges under dynamic conditions. The current flowing through the MOSFET terminals under the transient (dynamic) condition can be denoted as gate current, i_g; source current, i_s; drain current, i_d; and bulk current, i_b as illustrated in Figure 3.20 [31]. Moreover, their corresponding total charges to the terminal of MOSFET such as gate charges, drain charges, source charges, and bulk charges are denoted as Q_G, Q_D, Q_S, and Q_B, respectively. These charges are dependent on the terminal voltages of MOSFET and can be expressed as

$$Q_k = f\left(V_g, V_s, V_d, V_b\right) \; k = G, S, D, B \tag{3.26}$$

The total current flows on the terminal of MOSFET based on the Kirchhoff's law

$$i_g + i_s + i_d + i_b = 0 \tag{3.27}$$

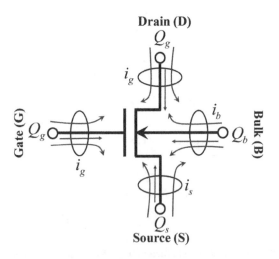

FIGURE 3.20   The flow of current through MOS transistor.

Using charge conservation law, the charges present can be obtained on the terminal as

$$Q_G + Q_S + Q_D + Q_B = 0 \qquad (3.28)$$

At any interval of time t, the charge present in the per unit area is the same as obtained by using DC voltage at the same time. This phenomenon is called quasi-state operation and the resulting dynamic model is called quasi-static model of MOSFET device.

The dynamic current expression can be obtained by summing the time-dependent transport current and charging current under the assumption of quasi-static operation as

$$i_s(t) = -i_s(V(t)) + \frac{dQ_s}{dt} \qquad (3.29)$$

$$i_d(t) = i_d(V(t)) + \frac{dQ_D}{dt} \qquad (3.30)$$

$$i_g(t) = \frac{dQ_G}{dt} \qquad (3.31)$$

$$i_b(t) = \frac{dQ_B}{dt} \qquad (3.32)$$

It should be noted that no current flows through the gate and substrate terminal, i.e., $I_g = I_b = 0$.

3.3.6 MEYER MODEL

In 1971, J. Meyer introduced a dynamic MOSFET model and it was extensively used as it did not have a charge conservation problem. In this model, the three lumped capacitances of gate channel are split as gate-to-source (C_{GS}), gate-to-drain (C_{GD}), and gate-to-bulk (C_{GB}) that can be defined as the derivative of charge at the gate with respect to source, drain, and bulk terminal, respectively [32, 33].

$$C_{GS} = \left.\frac{\partial Q_G}{\partial V_{gs}}\right|_{V_{gd},V_{gb}} , \quad C_{GD} = \left.\frac{\partial Q_G}{\partial V_{gd}}\right|_{V_{gs},V_{gb}} , \quad C_{GB} = \left.\frac{\partial Q_G}{\partial V_{gb}}\right|_{V_{gs},V_{gd}} \qquad (3.33)$$

where $V_{gd} = (V_{gs} - V_{ds})$ and $V_{gb} = (V_{gs} - V_{bs})$. The capacitance associated with the Meyer model is based on the following assumptions:

1. These capacitances of MOSFET are reciprocal.
2. Depending on the gate to bulk voltage (V_{gb}), the bulk charge density Q_B is constant along the channel length but independent of the source-to-drain voltage V_{ds}. This signifies that the bulk-to-source and drain (C_{BS} and C_{BD}) capacitances are zero.

Additionally, the total induced charge per unit area for NMOS transistor having inversion and depletion can be expressed using Gauss's law by assuming that the source and substrate are connected to the ground. In addition to that, the electron mobility is also assumed as a constant parameter in order to obtain the non-saturated drain current as

$$I_{ds} = \frac{W\mu_n c_1}{L}\left\{\left(V_{gs} - V_{fb} - 2\varphi_b - 0.5V_{ds}\right)V_{ds} - \frac{2\sqrt{2\varepsilon_s q N_a}}{3c_1}\left[\left(V_{ds} + 2\varphi_b\right)^{3/2} - \left(2\varphi_b\right)^{3/2}\right]\right\}$$

(3.34)

where c_1, φ_b, ε_s, N_a, and d_1 are denoted as the insulator capacitance per unit area $\left(c_1 = \varepsilon_1/d_1\right)$, energy separation, semiconductor permittivity, shallow acceptor density in P-type semiconductor, and insulator thickness, respectively. The drawback of the Meyer model is nonconservation of charge, which is important for the dynamic model. Therefore, the charge-based model needs to be considered for dynamic random access memory (DRAM) and switching capacitor filters.

3.4   MOSFET MODELING USING SPICE

In this section, the different levels of MOSFET models are used to obtain an accurate analysis based on the physical and electrical parameters.

3.4.1   BASIC CONCEPTS OF MODELING

In order to understand the experimental behavior of MOS transistor with the simulated result using industry-standard SPICE environments, this section describes the four different types of SPICE MOSFET model based on the physical aspects associated with each circuit level. This will help the reader to understand the difference between the various MOSFET models using SPICE. In this section, various built-in MOSFET models are described such as LEVEL-1, LEVEL-2, LEVEL-3, and LEVEL-4 (BSIM), etc. [34–37]. Figure 3.21 demonstrates the LEVEL-1 MOSFET equivalent circuit used in SPICE, where source-bulk and drain-bulk are represented by ideal diode that operates in reverse bias condition. Moreover, the source and drain resistance parasitics are represented by R_S and R_D, respectively.

3.4.2   MODEL EQUATIONS

In this subsection, the drain current equation is used to perform the circuit simulation under different levels of model. Moreover, their associated physical and electrical parameters have been considered to enhance the accuracy and performance of the various models along with the SPICE examples.

3.4.2.1   Level 1 Model Equation

In 1968, Shichman and Hodges introduced a Level 1 DC MOSFET model that is applicable for the device having gate length greater than 10 μm. The model equation is described by square-law current-voltage characteristics that are primarily used for

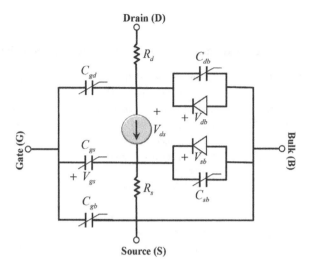

FIGURE 3.21   The equivalent circuit model of level 1 MOSFET.

longer-channel devices. This model provides an accurate analysis for long-channel devices based on Table 3.2. The drain current equation used in SPICE for the Level 1 NMOS transistor model is as follows:

In the case of the linear region:

$$I_{ds} = k'_n \left(\frac{W}{L_{eff}} \right) \left(V_{gs} - V_{th} - 0.5V_{ds} \right) V_{ds} \left(1 + \lambda V_{ds} \right) \ for \ V_{gs} \geq V_{th} \qquad (3.35)$$

In the case of saturation region:

$$I_{ds} = \frac{1}{2} k'_n \left(\frac{W}{L_{eff}} \right) \left(V_{gs} - V_{th} \right)^2 \left(1 + \lambda V_{ds} \right) \ for \ V_{gs} \geq V_{th} \qquad (3.36)$$

where $(1 + \lambda V_{ds})$ is defined as the empirical channel length modulation that can be observed in equations (3.35) and (3.36), although it is mostly observed in saturation region due to shorten physical channel length. The threshold voltage V_{th} and effective channel length L_{eff} can be expressed as

$$V_{th} = V_{t0} + \gamma \left(\sqrt{\left| 2\phi_f \right| + V_{sb}} - \sqrt{\left| 2\phi_f \right|} \right) \qquad (3.37)$$

where $L_{eff} = L - 2L_D$, $\gamma = \sqrt{2\varepsilon_{Si} q N_a} \ / \ C_{oxi}$, and $2\phi_f = 2\frac{kT}{q} \ln \left(\frac{n_i}{N_a} \right)$

Electrical parameters such as k', V_{t0}, γ, $|2\phi_f|$, and λ (represented in .model as KP, VT0, GAMMA, PHI, and LAMBDA, respectively) can be directly defined while using **.model** statement in the SPICE script file. Apart from electrical parameters,

TABLE 3.2
The Physical and Electrical Parameter of LEVEL 1 MOSFET SPICE Model

Parameters	SPICE Keyword	Parameter Description	Default Value	Units
V_{t0}	VTO	Zero-bias threshold voltage	0.0	V
K	KP	Transconductance parameter	2×10^{-5}	A/V^2
Γ	GAMMA	Body factor	0.0	
μ_0	UO	Low field mobility	600	cm^2/V s
$2\phi_f$	PHI	Surface potential in strong inversion	0.1	V
Λ	LAMBDA	Channel length modulation factor	0.0	V^{-1}
N_b	NSUB	Substrate doping	0.0	m^{-3}
t_{ox}	TOX	Gate oxide thickness	10^{-7}	m
N_{ss}	NSS	Surface state density	0.0	cm^{-2}
T_{pg}	TPG	Type of the gate material	1	–
I_s	IS	Bulk junction saturation current per sq-meter of the junction area	10^{-14}	A/m^2
J_s	JS	Bulk junction saturation current per sq-meter of the junction area	10^{-14}	A
R_s	RS	Source ohmic resistance	0.0	Ω
R_d	RD	Drain ohmic resistance	0.0	Ω
R_{sh}	RSH	Source, drain diffusion sheet resistance	Infinity	Ω/square
–	CBS	Zero-bias B-S junction capacitance	0.0	F
–	CBD	Zero-bias B-D junction capacitance	0.0	F
C_{j0}	CJ	Zero-bias bulk junction capacitance per sq-meter of the junction area	0.0	F/m^2
m_j	MJ	Bulk junction bottom grading coefficient	0.5	–
ϕ_{bi}	PB	Bulk junction potential	0.8	V
C_{jswo}	CJSW	Zero-bias bulk junction side-wall capacitance per meter of the junction perimeter	0.0	F/m
m_{jsw}	MJSW	Bulk junction sidewall grading coefficient	0.5	–
C_{gs0}	CGSO	Gate-source overlap capacitance per meter channel width	0.0	F/m
C_{gd0}	CGDO	Gate-drain overlap capacitance per meter channel width	0.0	F/m
C_{gb0}	CGBO	Gate-bulk overlap capacitance per meter channel length	0.0	F/m
–	KF	Flicker noise coefficient	0.0	–
–	AF	Flicker noise exponent	1	–

it is also possible to include the physical parameters into the **.model** statement such as μ, t_{ox}, and N_a that are referred to as mobility, oxide thickness, and P-type substrate doping concentration, respectively.

3.4.2.2 Level 2 Model Equation

In order to obtain a more precise model analysis, the Level 2 model considered a second-order effect for short-channel devices while obtaining the drain current. To enhance the analysis, it is also required to neglect the parameters that are assumed in the GCA analysis under the Level 1 model equation. In Level 2 model equation, the bulk depletion charge dependent on channel voltage is considered to obtain the drain current as

$$I_{ds} = \frac{k'}{(1-\lambda V_{ds})} \frac{W}{L_{eff}} \left\{ \left(V_{gs} - V_{fb} - |2\phi_f| - 0.5V_{ds} \right) V_{ds} - \frac{2}{3}\gamma \left[\left(V_{ds} - V_{bs} + |2\phi_f| \right)^{3/2} \right. \right.$$
$$\left. \left. - \left(-V_{bs} + |2\phi_f| \right)^{3/2} \right] \right\} \tag{3.38}$$

The channel length modulation is considered in equation (3.38) that also includes the variation of γ even though the V_{bs} becomes zero. Under saturation region, when the channel charges near the drain terminal become zero, the saturation voltage (V_{dsat}) and current (I_{dsat}) can be obtained as

$$V_{dsat} = V_{gs} - V_{fb} - |2\phi_f| + \gamma^2 \left(1 - \sqrt{1 + \frac{2}{\gamma^2}\left(V_{gs} - V_{fb} \right)} \right) \tag{3.39}$$

$$I_{ds} = I_{dsat} \frac{1}{(1-\lambda V_{ds})} \tag{3.40}$$

I_{dsat} can be obtained from equation (3.40) while substituting $V_{ds} = V_{dsat}$ in expression (3.39). The threshold voltage under zero bias supply can be obtained for the Level 2 model using equation (3.38) as

$$V_{t0} = \phi_{gc} - \frac{qN_{ss}}{C_{oxi}} + |2\phi_f| + \gamma\sqrt{|2\phi_f|} \tag{3.41}$$

where N_{ss} and ϕ_{gc} can be referred as work function difference of gate to the channel and fixed charge surface density, respectively. The performance of the Level 2 model is more accurate due to the additional physical parameters than the Level 1 model as summarized in Table 3.3. However, the accuracy of the Level 2 model is not sufficient to achieve a good agreement with the experimental data.

3.4.2.3 Level 3 Model Equation

The voltage and current equation of MOSFET is improved as presented in Level 2 for short channel down to 2 μm more precisely in Level 3. The improvement has been demonstrated using the Taylor series expansion of (3.38) under the linear region.

TABLE 3.3

The Parameters Associated with Level 2 in Addition to Table 3.2 (Level 1)

Parameters	SPICE Keyword	Parameter Description	Default Value	Units
L_{dif}	LD	Lateral diffusion	0.0	m
G_w	DELTA	Narrow width factor	0.0	–
X_j	XJ	Junction depth	0.0	m
u_l	UCRIT	Critical field for mobility degradation	1×10^4	V/cm
u_t	ULTRA	Mobility transverse field coefficient	0.0	–
V	UEXP	Exponent in mobility degradation	0.0	–
v_{max}	VMAX	Maximum carrier drift velocity	0.0	m/s
N_{eff}	NEFF	Effective substrate doping factor	1	–
N_{fs}	NFS	Fast surface state density	0.0	cm^{-2}
X_{qc}	XQC	Thin-gate oxide capacitance model flag and coefficient of channel charge share attributed to drain (0–0.5)	1.0	–

In order to improve the model, the semiempirical equation is used instead of an analytical equation by considering the short-channel effect. The short-channel effect is considered while calculating the threshold voltage and mobility in the empirical model. The drain current can be obtained in the linear region as

$$I_{ds} = \mu_s C_{oxi} \frac{W}{L_{eff}} \left(V_{gs} - V_{th} - \frac{1+F_b}{2} V_{ds} \right) V_{ds} \tag{3.42}$$

where $F_b = \frac{\gamma F_s}{4\sqrt{|2\phi_f| + V_{sb}}} + F_n$ is the empirical parameter that depends on the bulk depletion charge on the three-dimensional geometry of MOSFET. The short-channel effect based on the Dang's model is responsible to influence V_t, F_s, and μ_s, whereas the F_n is influenced by the narrow channel effects and μ_s is dependent on the gate electric field.

$$\mu_s = \frac{\mu}{1 + \theta\left(V_{gs} - V_{th}\right)} \tag{3.43}$$

In order to obtain more realistic values, the above expression can be rewritten as

$$\mu_{eff} = \frac{\mu_s}{1 + \mu_s \dfrac{V_{ds}}{v_{max} L_{eff}}} \tag{3.44}$$

Moreover, the threshold voltage can be expressed for SPICE Level 3 as

$$V_{th} = V_{t0} - \gamma\sqrt{|2\phi_f|} + \gamma F_s \sqrt{|2\phi_f| + V_{sb}} + F_n\left(|2\phi_f| + V_{sb}\right) - \sigma V_{ds} \tag{3.45}$$

where σ is the DIBL parameter and can be obtained using $\sigma = 8.15 \times 10^{-22} \eta / C_{oxi} L^3$. The physical and electrical parameters associated with the Level 3 MOSFET model are listed in Table 3.4.

TABLE 3.4

The Parameter Associated with Level 3 in Addition to Table 3.2 (Level 1)

Parameters	SPICE Keyword	Parameter Description	Default Value	Units
L_{dif}	LD	Lateral diffusion	0.0	m
G_w	DELTA	Narrow width factor	0.0	–
X_j	XJ	Junction depth	0.0	m
N_{fs}	NFS	Fast surface state density	0.0	cm^{-2}
Θ	THETA	Mobility degradation factor	0.0	V^{-1}
H	ETA	Static feedback factor	0.0	–
K	KAPPA	Saturation field correlation factor	0.2	–
v_{max}	VMAX	Maximum carrier drift velocity	–	–

3.4.2.4 BSIM Model

BSIM stands for the Berkeley Short Channel IGFET model that is the most commonly used model because it is more precise and highly accurate with the experimental data. The submicron technology was the first target by the UC Berkeley in the 1980s and 1990s that uses a feature size of 0.5 µm and 0.6 µm. Later, the binning process is introduced in BSIM3 to improve the accuracy of the DC and AC models. The binning process is demonstrated in Figure 3.22 that can vary foundry by foundry and process by process. The minimum and maximum width and channel length are depicted in Figure 3.22 with each bin. The parameters used in 0.18 µm and 0.13 µm technology are described in Table 3.5.

In the IC industry, the following different levels of MODEL associated with BSIM are used.

1. The first BSIM model used by the UC Berkley was Level 13.
2. The modified BSIM2 is Level 28 that uses 0.5 µm and 0.6 µm process technology.
3. The BSIM is a Level 49 model that is capable of modeling the circuit in the feature size below 0.1 µm. It is typically used in 0.35 µm, 0.25 µm, and 0.18 µm technology.
4. The BSIM is Level 54 capable of modeling the circuit in the feature size down to 0.13 µm process technology.

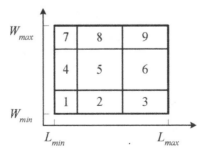

FIGURE 3.22   The concept of binning process.

TABLE 3.5

The Physical and Electrical Parameters Used in 0.18 μm and 0.13 μm Technology

Parameter	Symbol	0.18 μm nMOS	0.18 μm pMOS	0.13 μm nMOS	0.13 μm pMOS	Unit
Supply voltage	V_{dd}	1.8	1.8	1.2	1.2	V
Oxide thickness	t_{ox}	4.1	4.1	3.1	3.1	nm
Oxide capacitance	C_{oxi}	1.33	1.33	1.08	1.08	μF/cm²
Threshold voltage	V_{th}	0.37	-0.39	0.35	-0.35	V
Body effect coefficient	Γ	0.3	0.3	0.2	0.2	$V^{0.5}$
Fermi potential	$2|\phi_f|$	0.84	0.84	0.88	0.88	V
Junction capacitance	C_{j0}	0.95	1.17	0.95	1.15	fF/μm²
Built-in potential	ϕ_b	0.8	0.86	0.98	0.8	V
Grading coefficient	M	0.37	0.42	0.40	0.44	–
Nominal mobility	μ_0	292	112	430	100	cm²/V s
Effective mobility	μ_{eff}	287	88	298	97	cm²/V s
Saturation electric field	E_{sat}	1.2×10^5	2.5×10^5	9.5×10^4	4.2×10^5	V/cm
Equivalent on resistance	R_{eq}	8	22	18	37	kΩ/square
Saturation velocity	v_{sat}	8×10^6	8×10^6	1.5×10^7	1.5×10^7	cm/s

For a more detailed analysis of this model, the reader can understand by referring to the most recent literature and manual related to the BSIM model due to higher complexity.

3.4.3 Examples Using HSPICE

Example 3.1:

This HSPICE program demonstrates the voltage-transfer characteristic (VTC) curve for complementary metal oxide semiconductor (CMOS) inverter at 180 nm technology node based on the library considered from the predictive technology model.

SOLUTION:

In order to understand the VTC curve, the netlist is scripted below based on 180 nm technology node using industry-standard HSPICE simulation tool and the result is depicted in Figure 3.23.

```
Script Code:
*Based on 0.18 μm technology node, plot the VTC curve for CMOS
inverter
*In order to give comments on the script file, start the
statements with symbol '*'

******** Setting up various global parameters *********
```

```
* The .lib syntax is used to call the model library file saved in
the same location
.LIB "tsmc_018um_model.lib" CMOS_MODELS
.param Supply = 1.8V        * Supply voltage as Vdd globally Set
. opt scale = 0.09u         * Lambda value opted based on the rules


******** NMOS and PMOS Circuit description *************


M1   Vout    Vin    Gnd    Gnd    CMOSN    L = 2    W = 4
M2   Vout    Vin    Vdd    Vdd    CMOSP    L = 2    W = 8
Vdd  Vdd    Gnd    'Supply'
Vin     Vin    Gnd
******** *************
.dc      Vin    0    'Supply'     'Supply/100'    * It is used to
perform the DC analysis.


******** Plotting and printing statements **


.probe   V(Vout)
. option post=2
.end                    *In order to end the statement in a .sp file
```

RESULT:

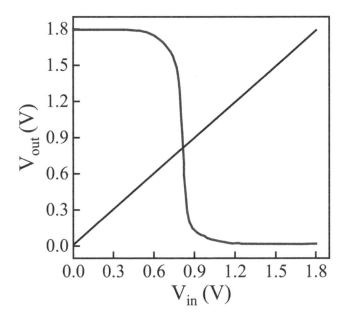

FIGURE 3.23 VTC of CMOS inverter at 180 nm technology.

Example 3.2:

This SPICE program demonstrates the variation of drain current as a function of gate voltages for the Level 1 DC MOSFET model.

SOLUTION:

In order to understand the *I–V* characteristics, the netlist is scripted below based on Level 1 MOSFET model using industry standard HSPICE simulation tool and the result is presented in Figure 3.24.

```
Script Code:
Vgs 1 0 2V
Vds 2 0 2V
Vbs 3 0 0V
*NMOS
M1 2 1 0 3 NCHL1 L=3u W=3u
*N-channel MOS Model
.MODEL NCHL1 NMOS LEVEL=1 RSH=0 TOX=300E-10 LD=0.21E-6 XJ=0.3E-6
+VMAX=15E4 ETA=0.18 GAMMA=0.4 KAPPA=0.5 NSUB=35E14 UO=700
+THETA=0.095 VTO=0.781 CGSO=2.8E-10 CGDO=2.8E-10 CJ=5.75E-5
CJSW=2.48E-10 +PB=0.7 MJ=0.5 MJSW=0.3 NFS=1E10
.options post probe
.option post=2
*For I-V Curve Id vs Vgs
.dc Vgs 0 5 0.1
.probe I1(M1)
*For I-V Curve Id vs Vds with Vgs = 2V
.dc vds 0 5 0.1
.probe I1(M1)
*For complete I-V Curve
.dc Vds 0 5 0.1 Vgs 0 5 1
.probe I1(M1)
.end
```

RESULT:

Variation of Id as a function of gate voltage in LEVEL 1 model

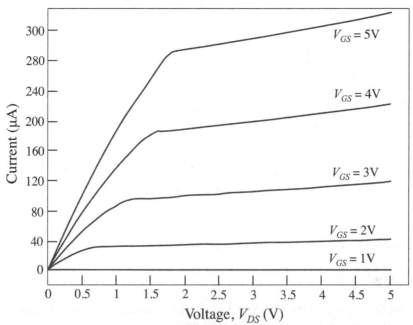

FIGURE 3.24   Drain current vs gate voltage characteristics in MOSFET DC LEVEL 1 model.

Example 3.3:

This SPICE program demonstrates the variation of drain current as a function of gate voltages for Level 2 DC MOSFET model.

SOLUTION:

In order to understand the I-V characteristics, the netlist is scripted below based on Level 2 MOSFET model using industry standard HSPICE simulation tool and the resultant plot is demonstrated in Figure 3.25.

```
Script Code:
Vgs 1 0 2V
Vds 2 0 2V
Vbs 3 0 0V
*NFET
M1 2 1 0 3 NCHL2 L=3u W=3u
*N-channel MOS Model
.MODEL NCHL2 NMOS LEVEL=2 PHI=0.7 TOX=4.08E-8 XJ=0.2 TPG=1
VTO=0.8096 +DELTA=4.586 LD=2.972E-7 KP=5.3532E-5 UO=632.5
UEXP=0.1328 UCRIT=3.897E4 +RSH=6.202 GAMMA=0.5263 NSUB=5.977E15
NFS=5.925e11 VMAX=5.83E4 +LAMBDA=3.903E-2 CGDO=3.7731E-10
CGSO=3.7731E-10 CGBO=3.4581E-10 +CJ=1.3679E-4 MJ=0.63238
CJSW=5.1553E-10 MJSW=0.26805 PB=0.4
.options post probe
*For I-V Curve Id vs Vgs
```

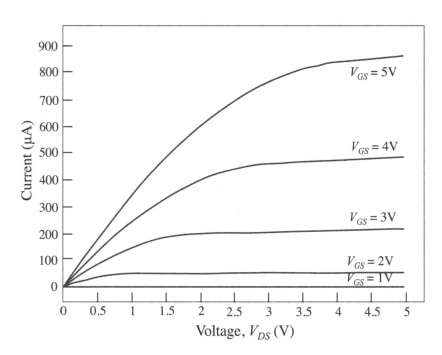

FIGURE 3.25 Drain current vs gate voltage characteristics in MOSFET DC LEVEL 2 model.

```
.dc Vgs 0 5 0.1
.probe I1(M1)
*For I-V Curve Id vs Vds with Vgs = 2V
.dc vds 0 5 0.1
.probe I1(M1)
*For complete I-V Curves
.dc Vds 0 5 0.1 Vgs 0 5 1
.probe I1(M1)
.end
```

RESULT:

Variation of Id as a function of gate voltage in LEVEL 2 model

Example 3.4:

This SPICE program demonstrates the variation of drain current as a function of gate voltages for Level 3 DC MOSFET model.

SOLUTION:

In order to understand the I-V characteristics, the netlist is scripted below based on Level 3 MOSFET model using industry-standard HSPICE simulation tool and the result is shown in Figure 3.26.

```
Script Code:
Vgs 1 0 2V
Vds 2 0 2V
Vbs 3 0 0V
*N-channel MOS model
M1 2 1 0 3 NCHL3 L=3u W=3u
*NFET Model
.MODEL NCHL3 NMOS LEVEL=3 TPG=1 TOX=1.5E-8 LD=2.95E-7 WD=3.00E-7
UO= 263 +VTO=0.5 THETA=0.046 RS=27 RD=27 DELTA=2.27 NSUB=1.45E17
XJ=1.84E-7 +VMAX=1.10E7 ETA=0.927 KAPPA=0.655 NFS=3E11
CGSO=3.4E-10 CGDO=3.48E-10 +CGBO=5.75E-10 XQC=0.4
.options post probe
*For I-V Curve Id vs Vgs
.dc Vgs 0 5 0.1
.probe I1(M1)
*For I-V Curve Id vs Vds with Vgs = 2V
.dc vds 0 5 0.1
.probe I1(M1)
*For complete I-V Curves
.dc Vds 0 5 0.1 Vgs 0 5 1
.probe I1(M1)
.end
```

RESULT:

Variation of Id as a function of gate voltage in LEVEL 3 model

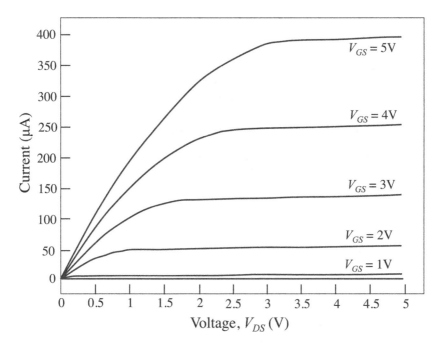

FIGURE 3.26   Drain current vs gate voltage characteristics in MOSFET DC LEVEL 3 model.

3.5   SUMMARY

This chapter can be summarized as follows:

- This chapter briefly explained the basic structure of MOSFET along with its characteristics and importance of scaling. However, the impact of scaling was also observed due to the hot carrier effect, short-channel effect, etc. Further, the importance of the MOS capacitor device was demonstrated and characterized by their $C–V$ curve under different regions.
- Moreover, the different DC MOSFET models were demonstrated for the static and dynamic condition under different operating regions by obtaining the drain current expression.
- Based on the MOSFET model, circuit modeling was examined for different levels of the model in order to validate experimental data with the simulation result in a more precise way with higher-order parameters.
- We concluded this chapter with a few SPICE netlist related to the different levels of the MOSFET model.

3.6   MULTIPLE-CHOICE QUESTIONS

1. The depletion N-channel MOSFET
   a. Can be operated as a junction field-effect transistor (JFET) with zero gate voltage
   b. Can be operated as an enhancement MOSFET by applying +ve bias to gate

    c.   Can be operated as an enhancement MOSFET by applying −ve bias to gate

    d.   Cannot be operated as an enhancement MOSFET

2. Choose the correct statement:
    a.   MOSFET is an uncontrolled device
    b.   MOSFET is a current-controlled device
    c.   MOSFET is a voltage-controlled device
    d.   MOSFET is a temperature-controlled device

3. The arrow on the symbol of MOSFET indicates
    a.   that it is an N-channel MOSFET
    b.   the direction of electrons
    c.   the direction of conventional current flow
    d.   that it is a P-channel MOSFET

4. The controlling parameter in MOSFET is
    a. V_{ds}
    b. I_g
    c. V_{gs}
    d. I_s

5. The output characteristics of a MOSFET is a plot of
    a. I_d as a function of V_{gs} with V_{ds} as a parameter
    b. I_g as a function of V_{gs} with V_{ds} as a parameter
    c. I_d as a function of V_{ds} with V_{gs} as a parameter
    d. I_g as a function of V_{ds} with V_{gs} as a parameter

6. Consider an ideal MOSFET. If $V_{gs} = 0V$, then $I_d =$?
    a.   Zero
    b.   Maximum
    c. I_d (ON)
    d. I_{dd}

7. The transistor can be operated in
    a.   active region
    b.   saturation region
    c.   cut-off region.
    d.   all of the above regions.

8. The main advantage of short-channel devices is _____.
    a.   its power consumption is low
    b.   it has good output characteristics
    c.   it has high speed
    d.   it is easy to fabricate

9. Subthreshold operation occurs in _____.
    a.   strong inversion region
    b.   weak inversion

   c. saturation region

   d. cut-off region

10. Which one is not a second-order effect?
   a. Body effect
   b. Channel length modulation
   c. Subthreshold conduction
   d. Hot carrier effect

3.7 SHORT ANSWER QUESTIONS

1. Define the strong inversion layer and the threshold voltage of a MOS system.
2. What are the three regions of operation in the MOS transistor?
3. What is the gradual-channel approximation (GCA) model?
4. What is the difference between constant field scaling and constant voltage scaling?
5. What do you mean by hot carrier effect?
6. What is the meaning of body effect?
7. What is the meaning of the short-channel effect?
8. Explain the phenomenon of DIBL.
9. What factors may affect the threshold voltages?
10. What are the second-order effect?

3.8 LONG ANSWER QUESTIONS

1. What do you mean by drain-induced barrier lowering (DIBL)?
2. Derive an expression for saturated drain current considering channel length modulation.
3. What are the short-channel effects? Discuss them in detail.
4. What are the different MOSFET capacitances? Discuss each of them with their origins.
5. Draw and explain the MOS $C–V$ characteristic.
6. Discuss the MOSFET modeling with level-1 model parameters.
7. Discuss the MOSFET modeling with level-2 model parameters.
8. Discuss the MOSFET modeling with level-3 model parameters.
9. Discuss the BSIM model.

REFERENCES

1. Faggin, F. and Klein, T. 1970. Silicon gate technology. *Solid-State Electronics* 13, 1125–1144.
2. Appels, J. A., Kooi, E., Paffen, M. M., Schiorje, J. J. H. and Verkuylen, W. H. C. G. 1970. Local oxidation of silicon and its application in semiconductor technology. *Philips Research Reports* 25, 118–132.
3. Sah, C. T. 1964. Characteristics of the metal-oxide-semiconductor transistors. *IEEE Transactions on Electron Devices* 11, no. 7: 324–345.

4. Nicollian, E. H. and Brews, J. R.. 1982. *MOS (Metal Oxide Semiconductor) Physics and Technology*, John Wiley & Sons, New York, NY.

5. Tsividis, Y. P. 1987. *Operation and Modeling of the MOS Transistor*, McGraw-Hill Book Company, New York, NY.

6. Hsu, F. C., Ko, P. K., Tam, S., Hu, C. and Muller, R. S. 1982. An analytical breakdown model for short-channel MOSFETs. *IEEE Transactions on Electron Devices* 29, no. 11: 1735–1740.

7. Hsu, F. C., Muller, R. S. and Hu, C. 1983. A simplified model of short-channel MOSFET characteristics in the breakdown mode. *IEEE Transactions on Electron Devices* 30, no. 6: 571–576.

8. Takeda, E. 1984. Hot-carrier effects in submicrometer MOS VLSIs. In *Proceeding of IEEE Solid-State and Electron Devices* 131, no. 5: 153–164.

9. Ng, K. K. and Lynch, W. T. 1987. The impact of intrinsic series resistance as MOSFET scaling. *IEEE Transactions on Electron Devices* 34, no. 3: 503–511.

10. Gildenblat, G. S. and Cohen, S. S. 1987. Contact metallization. In: N. G. Einspruch and G. Gildenblat (Eds.), *VLSI Electronics* 15, Academic Press Inc., New York, NY.

11. Pimbley, M., Cumberbatch, E. and Hagan, P. S. 1987. Analytical treatment of MOSFET source-drain resistance. *IEEE Transactions on Electron Devices* 34, no. 4: 834–838.

12. Baccarani, G. and Sai-Halasz, G. A. 1983. Spreading resistance in submicron MOSFETs. *IEEE Transactions on Electron Device Letters* 4, no. 2: 27–29.

13. Ng, K. K. and Lynch, W. T. 1986. Analysis of the gate-voltage dependent series resistance of MOSFETs. *IEEE Transactions on Electron Devices* 33, no. 7: 965–972.

14. Chou, S. Y. and Antoniadis, D. A. 1987. Relationship between measured and intrinsic transconductances of FETs. *IEEE Transactions on Electron Devices* 34, no. 2: 448–450.

15. Cserveny, S. 1990. Relationship between measured and intrinsic transconductances of MOSFETs. *IEEE Transactions on Electron Devices* 37, no. 11: 2413–2414.

16. Seavey, M. H. 1984. Source and drain resistance determination for MOSFETs. *IEEE Electron Device Letters* 5, no. 11: 479–481.

17. Liu, S. and Nagel, L. W. 1982. Small-signal MOSFET models for analog circuit design. *IEEE Journal of Solid-State Circuits* 17, no. 6: 983–998.

18. Huang, C. L. and Gildenblat, G. Sh. 1989. MOS flat-band capacitance method at low temperatures. *IEEE Transactions on Electron Devices* 36, no. 8: 1434–1439.

19. Werner, W. M. 1974. The work function difference of MOS system with aluminum field plates and polycrystalline silicon field plates. *Solid-State Electronics* 17, no. 8: 769–775.

20. Stern, F. and Howard, W. E. Properties of semiconductor surface inversion layers in quantum limits. *Physical Review* 163, no. 3: 816–835.

21. Zaininger, K. H. and Heiman, F. 1970. The C–V technique as an analytical tool. *Solid-State Technology* Part I: 49–56; Part II: 46–55.

22. Berman, A. and Kerr, D. R. 1974. Inversion charge distribution model of the high frequency MOS capacitance. *Solid-State Electronics* 17, no. 7: 735–742.

23. Yaron, G. and Frohman-Bentchkowsky, D. 1980. Capacitance voltage characterization of poly Si-SiO$_2$-Si structures. *Solid-State Electronics* 23, 433–439.

24. Jaeger, R. C., Gaensslen, F. H. and Diehl, S. E. 1983. An efficient algorithm for simulation of MOS capacitance. *IEEE Transactions on Computer-Aided Design* 2, no. 2: 111–116.

25. Pao, H. C. and Sah, C. T. 1966. Effects of diffusion current on characteristics of metal-oxide (insulator)-semiconductor transistors. *Solid-State Electronics* 9, no. 10: 927–937.

26. Brews, R. 1978. A charge sheet model of the MOSFET. *Solid-State Electronics* 21, no. 2: 345–355.

27. Wilson, C. L. and Blue, J. L. 1982. Two-dimensional finite element charge-sheet model of a short channel MOS transistor. *Solid-State Electronics* 25, no. 6: 461–477.

28. Brews, R. 1981. Physics of MOS transistor. In Kahng, D. (Ed.), *Silicon Integrated Circuits, Part A*, Applied Solid-State Science Series, Academic Press, New York, NY.

29. Guebels, P. P. and Wiele, F. V. 1983. A small geometry MOSFET models for CAD applications. *Solid-State Electronics* 26, no. 4: 267–273.

30. Kotani, N. and Kawazu, S. 1979. Computer analysis of punch-through in MOSFETs. *Solid-State Electronics* 22, no. 1: 63–70.

31. Ward, D. and Dutton, R. W. 1978. A charge-oriented model for MOS transistor capacitances. *IEEE Journal of Solid-State Circuits* 13, no. 5: 703–707.

32. Meyer, J. 1971. MOS models and circuit simulation. *RCA Review* 32: 42–63.

33. Cirit, M. A. 1989. The Meyer model revisited: Why is charge not conserved? *IEEE Transactions on Computer-Aided Design* 8, no. 10: 1033–1037.

34. Vladimirescu, A. and Liu, S. 1980. The simulation of MOS integrated circuits using SPICE2. *Memorandum No. UCBjERL M80/7*, Electronics Research Laboratory, University of California, Berkeley.

35. Sheu, B., Scharfetter, D., Ko, P. and Jeng, M. 1987. BSIM: Berkeley short-channel IGFET model for MOS transistors. *IEEE Journal of Solid-State Circuits* 22, no. 4: 558–566.

36. Ward, D. E. 1982. Charge-based modeling of capacitances in MOS transistors. *Stanford University Technical Report*: 201–211.

37. Hsu, M. G. and Sheu, B. J. 1987. Inverse-geometry dependence of MOS transistor electrical parameters. *IEEE Transactions on Computer-Aided Design* 6, no. 4: 582–585.

4 Combinational and Sequential Design in CMOS

4.1 CMOS INVERTER

Complementary metal oxide semiconductor (CMOS) inverter is a micro and vital element of the very-large-scale integration (VLSI) circuit design wherein the analysis makes the VLSI engineer aware about the several parameters of technology. By analyzing the inverter design, the designer can estimate the power, energy, and area overheads of the VLSI system.

4.1.1 Design

The CMOS inverter basically covers the features of both *P*-channel metal oxide semiconductor (PMOS) and NMOS inverters; circuit diagram is shown in Figure 4.1. In a CMOS inverter, a PMOS transistor is placed between the supply voltage and output, and the NMOS transistor is placed between the output and the ground terminal; this is because PMOS and NMOS transistors are responsible to produce full logic "1" and full logic "0", respectively. The gates of both PMOS and NMOS are tied together to form an input and the drains are tied to form an output of the inverter [1]. To avoid the body biasing effect and leakages, the body terminal of the PMOS and NMOS is connected to the respective source terminal of the transistor, as shown in Figure 4.1.

4.1.2 Operation

The functionality of the CMOS inverter is illustrated using switching diagrams (see Figure 4.1). When the input is very low, the PMOS transistor is switched ON and shows lower ON resistance and the NMOS transistor is switched OFF and demonstrates high resistance. As a result, the load capacitance connected at the output charges to V_{DD} through the PMOS transistor. Similarly, when the applied input is very high, the NMOS transistor is switched ON and shows low ON resistance and the PMOS transistor is switched OFF and shows high resistance. As a result, the load capacitance at the output discharges to the ground through the NMOS transistor. Thus, the designed circuit with PMOS and NMOS transistors acts as an inverter, commonly known as the NOT gate.

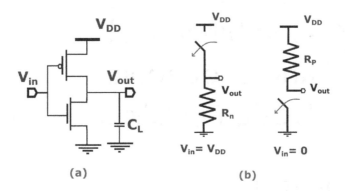

FIGURE 4.1   (a) CMOS inverter design. (b) Switching diagram of a CMOS inverter.

4.1.3 TRANSIENT AND VTC CHARACTERISTICS

When a pulse signal with a frequency (f) is applied to the CMOS inverter design, the output is an inverted signal, as shown in Figure 4.2(a). From the output signal, it can be observed that the logic levels of the output are fully restored. The reason behind this full logic swing is explained in the following sections. Similarly, Figure 4.2(b) shows the voltage transfer characteristics (VTC) of the CMOS inverter with V_{in} on the x-axis and V_{out} on the y-axis [1]. From Figure 4.2(b), it can be observed that at different regions of the VTC curve, the PMOS and NMOS operate in different regions.

4.1.4 SIGNIFICANCE OF THE CMOS INVERTER

CMOS explores both NMOS and PMOS characteristics and achieves several benefits that can be explained using the following metrics [1, 2]:

High logic levels: With the use of both PMOS and NMOS designs in the inverter, it achieves high logic swing, that is, the inverter restores the full logic levels. As a result, the noise margin of the logic gates is high.

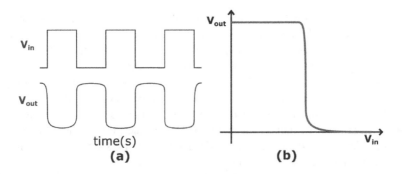

FIGURE 4.2   CMOS inverter (a) Transient characteristics (b) Voltage transfer characteristics.

TABLE 4.1

Performance Comparison of the CMOS Inverter with the NMOS Inverter

Property	CMOS Inverter	NMOS Inverter
Logic Swing	Restored logic swings	Low logic swing
Power consumption	Low	High
Noise Margin	High	Low
Logic	Ratio less logic	Ratio logic
Transistors used	Both PMOS and NMOS	Only NMOS

   Low power consumption: In steady state, there is no path between the supply
      voltage and the ground of the CMOS inverter. Since only one transistor is
      switched ON at a time, the CMOS inverter achieves ultra-low static power
      consumption.
   Ratioless design: In the CMOS inverter, the obtained logic levels are inde-
      pendent of device sizes. This makes the designing of CMOS-based circuits
      much simpler.
   High input resistance: In steady state, due to the insulated gate terminal,
      no current enters into the gate terminal. As a result, the CMOS inverter
      achieves high fan-out (ideally infinite).

The summary of the properties and benchmarking of the CMOS inverter against
the NMOS inverter is listed in Table 4.1. The detailed description of the above-
mentioned benefits and properties is given in the subsequent sections of the chapter.

4.2   STATIC BEHAVIOR OF THE INVERTER

To study the static behavior of the CMOS inverter and to explain the robustness of
the inverter design, several parameters are evaluated, such as switching threshold,
noise margin, and minimum supply voltage. These parameters are derived by utilizing the voltage transfer characteristics of the CMOS inverter.

4.2.1   SWITCHING THRESHOLD

Switching threshold (V_M) of a CMOS inverter is a point on the voltage transfer characteristics curve, where V_{in} and V_{out} of the CMOS inverter are equal. On the voltage
transfer characteristics curve, this value is marked as the intersection of the voltage
transfer characteristics with $V_{in} = V_{out}$ line, as shown in Figure 4.3. As V_{GS} is equal
to V_{DS}, at this point, both PMOS and NMOS transistors operate in the saturation
region [3].

   V_M is mathematically expressed as equation (4.1).

$$V_M = \left(\frac{rV_{DD}}{(1+r)} \right) \tag{4.1}$$

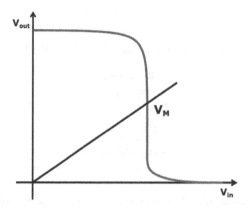

FIGURE 4.3 Voltage transfer characteristics indicating switching threshold.

The value of V_M is determined by using the value "r" that is obtained by using the sizing of both PMOS and NMOS. The ideal value of V_M is always $V_{DD}/2$ and has a high noise margin in the middle of the voltage transfer characteristics. To move V_M to the upper side, the PMOS device should be wider in size. Similarly, to move to the lower side, the NMOS size should be wider.

4.2.2 NOISE MARGIN

High-level (NM_H) and low-level noise margins (NM_L) of the CMOS inverter are defined as equations (4.2) and (4.3).

$$NM_H = V_{OH} - V_{IH},\qquad(4.2)$$

$$NM_L = V_{IL} - V_{OL},\qquad(4.3)$$

where V_{OH} is the maximum output voltage of CMOS inverter with high-level input V_{IH} and V_{OL} is the minimum output voltage of the CMOS inverter with low-level input V_{IL}.

V_{IL} and V_{IH} values are calculated at the gain of the amplifier equal to −1. The simplistic approach to identify these values is to derive the voltage transfer characteristics and locate the values of V_{IL} and V_{IH}, as shown in Figure 4.4.

4.2.3 ROBUSTNESS OF THE CMOS INVERTER BY SCALING SUPPLY VOLTAGE

Robustness of the device is evaluated by scaling the supply voltage of the inverter design. This analysis results lower limit of the supply voltage with technology scaling. This can be determined by observing the voltage transfer characteristics with the supply voltage scaling, as shown in Figure 4.5. From this, it can be observed that the inverter shows acceptable characteristics for supply voltage equal to the threshold voltage of the transistor. With the supply voltages lower than the

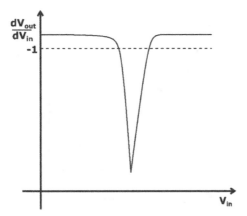

FIGURE 4.4   CMOS inverter voltage gain versus input voltage.

threshold voltages, the performance of the inverter design is degraded, as shown in Figure 4.5. It has been predicted that in order to have a proper operation, the supply voltage of the inverter is greater than twice the threshold voltage of the transistor [1–5].

4.3   DYNAMIC BEHAVIOR OF CMOS INVERTER

Dynamic behavior of the CMOS inverter presents the propagation delay calculation by considering different capacitive components. First, the different capacitive components of the CMOS inverter are described. Later, the propagation delay of the CMOS inverter is calculated.

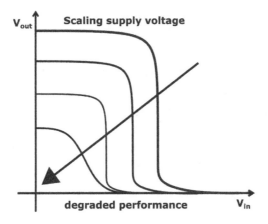

FIGURE 4.5   VTC of CMOS inverter with scaling supply voltage.

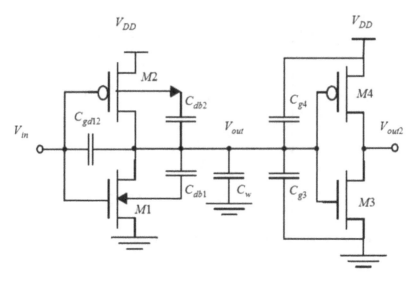

FIGURE 4.6 Parasitic capacitances, influencing the transient behavior of the cascaded inverter pair.

4.3.1 CAPACITANCES

The inverter is loaded with several capacitive components, and to calculate the delay of the inverter, all these components need to be considered. Figure 4.6 shows the several capacitive components that influence the behavior of the CMOS inverter.

4.3.1.1 Gate-Drain Capacitance

This capacitance comprises the overlap capacitance of both transistors M_1 and M_2. This capacitance is the overlap capacitance between gate-drain terminals of both transistors. This capacitance is expressed as

$$C_{gd} = 2\ C_{GD0}W,\qquad(4.4)$$

where W is the transistor width and C_{GD0} is the zero bias gate to drain capacitance.

4.3.1.2 Diffusion Capacitance (C_{db1}, C_{db2})

This capacitance is the depletion capacitance of the reverse-biased *PN* diode of the drain to body junction. This capacitor is nonlinear in nature and highly dependent on the applied voltage. This can be expressed as

$$C_{db} = K\ C_{j0},\qquad(4.5)$$

where K is the constant, and C_{j0} is the junction capacitance.

4.3.1.3 Gate Capacitance (C_{g3}, C_{g4})

Gate capacitance mainly consists of oxide capacitance and depends on the length and width of the transistor. It can be expressed as

$$C_g = W_n L_n C_{ox} + W_p L_p C_{ox},$$ (4.6)

where W_n and W_p are the widths of NMOS and PMOS, respectively.

L_n and L_p are the lengths of NMOS and PMOS, respectively.
C_{ox} is the oxide capacitance.

4.3.1.4 Propagation Delay of the CMOS Inverter

The propagation delay of the inverter that charges and discharges the load capacitor is expressed as

$$t_p = \int_{v_1}^{v_2} \frac{C_L(v)}{i(v)} dv,$$ (4.7)

where $i(v)$ and $C_L(v)$ are the charge currents and load capacitance as a function of voltage, respectively. Now, consider the simplified switch model of the CMOS inverter shown in Figure 4.7 that replaces the ON resistance value, as shown in equation (4.8).

$$R_{eq} = \frac{1}{V_{DD}/2} \int_{V_{DD}/2}^{V_{DD}} \frac{V}{I_{DSAT}(1+\lambda V)} dV \approx \frac{3}{4} \frac{V_{DD}}{I_{DSAT}} \left(1 - \frac{7}{9}\lambda V_{DD}\right)$$ (4.8)

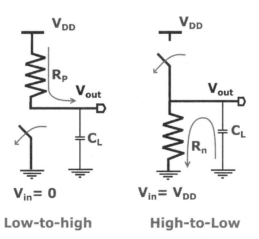

Low-to-high   High-to-Low

FIGURE 4.7   Switch model of a CMOS inverter.

with

$$I_{DSAT} = k\frac{W}{L}\left((V_{DD} - V_T)V_{DSAT} - \frac{V_{DSAT}^2}{2}\right)$$

It is a fact that deriving the propagation delay of the first-order RC network shown in Figure 4.7 is always clear from the basics and it is expressed as

$$t_{pHL} = \ln(2)R_{eqn}C_L = 0.69R_{eqn}C_L \qquad (4.9)$$

Propagation delay of high-to-low transition and low-to-high transition is expressed as

$$t_{pLH} = 0.69R_{eqn}C_L, \qquad (4.10)$$

where R_{peq} and R_{neq} are the equivalent ON resistance of both PMOS and NMOS, respectively, and C_L is the load capacitance. The overall propagation delay of the inverter is the average value of t_{pLH} and t_{pHL} and can be expressed as

$$t_p = \frac{t_{pHL} + t_{pLH}}{2} = 0.69C_L\left(\frac{R_{eqn} + R_{eqp}}{2}\right) \qquad (4.11)$$

4.3.2 POWER AND ENERGY CONSUMPTION

This section presents several different components of the CMOS inverter power consumption such as static and dynamic power consumption. Further, it also explains the energy consumption and the energy delay of the CMOS inverter.

4.3.2.1 Power Consumption

The power consumption of the CMOS inverter mainly comprises three components: dynamic power consumption, static power consumption, and direct path power consumption.

4.3.2.2 Dynamic Power Consumption

This component of power consumption is the significant part of the total power consumption of the CMOS inverter [6]. Dynamic power dissipation in the CMOS inverter is mainly due to the charging and discharging of the load capacitor. During low-to-high transition, the PMOS device charges the capacitor wherein the capacitor draws some energy from the power supply. Later, the capacitor charges to V_{DD} and in high-to-low transition, this energy is dissipated through the NMOS transistor. The energy consumption can be derived by considering low-to-high transition, as shown in Figure 4.8. The value of energy consumption drawn from the supply voltage is expressed in equation (4.12).

$$E_{VDD} = \int_0^\infty i(t)V_{DD}dt = V_{DD}\int_0^\infty C_L\frac{dv_{out}}{dt}dt = C_LV_{DD}\int_0^{V_{DD}} dv_{out} = C_LV_{DD}^2 \qquad (4.12)$$

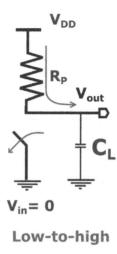

Low-to-high

FIGURE 4.8 Switch model of CMOS inverter in low-to-high transition.

On the other hand, the energy stored in the capacitor is $C_L V_{DD}^2/2$, which means the capacitor stores only half of the energy drawn from the supply voltage. The remaining half of the energy is dissipated by the transistor and this dissipation is independent of the transistor size. Now, the power consumption of the CMOS inverter is defined in equation (4.13).

$$P_{dyn} = C_L V_{DD}^2 f_{0-1} \qquad (4.13)$$

where f_{0-1} represents the frequency of energy consumption of the transitions, that is, how often the device draws the energy from the supply voltage and the transistor consumes it. The dynamic power consumption by using the switching activity factor is shown in equation (4.14).

$$P_{dyn} = C_L V_{DD}^2 f_{0-1} = C_L V_{DD}^2 p_{0-1} f, \qquad (4.14)$$

where p_{0-1} represents the probability of low-to-high switching phenomenon.

4.3.2.3 Static Power Consumption

Static power consumption or steady-state power consumption of the CMOS inverter is ideally zero. Since the PMOS and NMOS are never switched ON simultaneously in the steady state, this power consumption is due to the leakage of the reverse-biased diodes. The leakage is primarily due to the drain or source and body junctions in the CMOS inverter, as shown in Figure 4.9, and expressed in equation (4.15).

$$P_{stat} = I_{stat} V_{DD} \qquad (4.15)$$

where I_{stat} is the current flow in the steady state, generally it is very low in magnitude. This static power consumption increases with reduction in the channel length of the transistor.

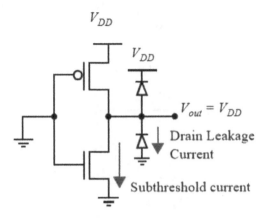

FIGURE 4.9   Source of leakage currents in a CMOS inverter.

4.3.2.4   Direct Path Power Consumption

This power consumption occurs due to the finite slope of the input signal applied. Owing to this finite slope, there exists a path between V_{DD} and ground (GND) for a fraction of the time period. This causes a peak current (I_{peak}) for a fraction of time, as shown in Figure 4.10. This is called short circuit or direct path power consumption and it is expressed in equation (4.16).

$$P_{dp} = t_{sc}V_{DD}I_{peak}f = C_{sc}V_{DD}{}^2 f \tag{4.16}$$

4.3.2.5   Total Power Consumption

Putting all these individual power consumption components together makes the overall power consumption of the CMOS inverter, it is expressed in equation (4.17).

$$P_{tot} = P_{dyn} + P_{dp} + P_{stat} = \left(C_L V_{DD}{}^2 + V_{DD}I_{peak}t_s\right)f_{0-1} + V_{DD}I_{leak} \tag{4.17}$$

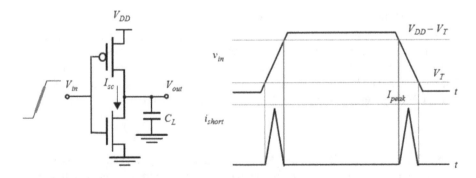

FIGURE 4.10   Short circuit current with input transitions.

Example 4.1:

Calculate the capacitive dissipation and energy of the CMOS inverter, where the value of the load capacitance was found to be to equal 6 fF. For a supply voltage of 2.5 V, the CMOS inverter shows a propagation delay of 32.5 psec.

ANSWER:

$$E_{dyn} = C_L V_{DD}^2 = 37.5 \text{ fJ}$$

Assume that the inverter is operating at the maximum possible rate ($T = 1/f = t_{pLH} + t_{pHL} = 2t_p$). For a t_p of 32.5 psec, we find that the dynamic power dissipation of the circuit is

$$P_{dyn} = E_{dyn} / (2t_p) = 580 \text{ μW}$$

4.4   DESIGN OF COMBINATIONAL LOGIC DESIGN

This section presents the designing of combinational circuits and deals with different topologies. The previous section dealt with the CMOS inverter whereas in this section, two and multiple input logic gates have been designed. Further, the advantages and disadvantages of these topologies have been highlighted and benchmarked.

4.4.1   COMPLEMENTARY CMOS LOGIC

In general, a static CMOS circuit is built using combinations of two networks. The upper part, which is connected to the supply voltage (V_{DD}) is called a pull-up network (PUN) and the part of the network connected to GND is called a pull-down network (PDN), as shown in Figure 4.11. The PUN connects the output and

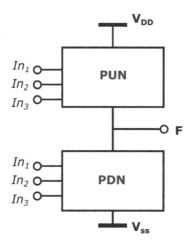

FIGURE 4.11   Block diagram of complementary CMOS logic.

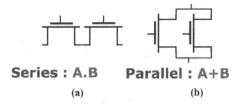

Series : A.B   Parallel : A+B

(a)   (b)

FIGURE 4.12   PDN NMOS rules [5].

the V_{DD} and the PDN connects the output and the GND. The pull-up and pull-down networks are designed using PMOS and NMOS, respectively. The pull-up and pull-down networks are designed in such a way that in the steady state, only one network conducts [4, 6].

4.4.1.1   Guidelines in Designing Static CMOS Logic

Detailed guidelines are as follows:

- NMOS switches ON with a higher gate voltage and switches OFF with a lower gate voltage. On the other hand, PMOS switches ON with a lower gate voltage and switches OFF with a higher gate voltage.
- PUN consists of PMOS transistors and PDN consists of NMOS transistors. The reason behind this selection is that NMOS produces strong 0 and PMOS produce strong 1.
- The fundamental rule in designing static CMOS circuits is that in PDN, the NMOS devices are connected in series resulting in an AND logic, as shown in Figure 4.12(a). Similarly, NMOS devices are connected in parallel resulting in an OR logic, as shown in Figure 4.12(b).
- It can be observed that the PUN and PDN are dual in nature. That is, every parallel combination of devices in PUN consists of a series combination of the devices in PDN and vice versa.
- Every M-input logic circuit utilizes 2M number of transistors to implement. The complementary logic is inverting logic wherein these gates in a single stage are able to realize logic gates like NAND, NOR, and XNOR.

4.4.1.2   Two- and Multi-Input Static Complementary Gates

Figure 4.13 shows the static complementary two-input NAND gate. The PDN consists of two NMOS transistors that are connected in series. In dual, the PUN consists of two PMOS transistors, which are connected in parallel. When A and B are both zero, two NMOS transistors are switched OFF, PMOS transistors are switched ON, and the output is connected to the supply voltage (logic 1). It can be concluded that the output (F) is connected either to V_{DD} or GND, but not to both simultaneously.

Designing of complex logic gates is explained as follows, consider an expression

$$Y = \overline{(D + A. (B+C))}$$

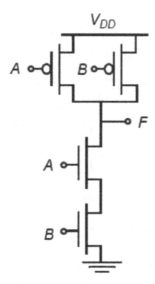

FIGURE 4.13   Static complementary NAND gate.

The first step is the designing of the PDN, following the fact that in PDN, NMOS in series forms the AND logic or NMOS in parallel forms the OR logic. The expression is divided into subcircuits and the implementation is performed. B and C are performing OR operation and are connected in parallel. Later, A is connected in series with B and C combination. Finally, D is connected in parallel to the designed circuit and this completes the PDN, as shown in Figure 4.14. The dual of this is implemented in PUN with PMOS circuits (Figure 4.14 shows the complete circuit diagram).

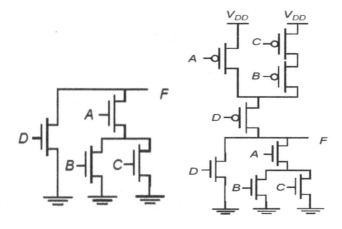

FIGURE 4.14   Static complementary design of complex design.

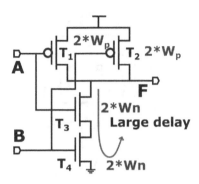

FIGURE 4.15 Sizing CMOS NAND gate for optimum delay.

4.4.1.3   Sizing Static Complementary Gates for optimum Propagation Delay

Consider a two-input NAND gate and its *RC* switch model. From this switching model, it can be understood that the propagation delay depends on the input patterns applied. For example, if both inputs are low, the propagation delay will be $0.69 \times (R_p/2) \times C_L$ since both the PMOS devices are in parallel. This is not the case when one input is logic 0 and other is logic 1. In this case, the delay is $0.69 \times R_p \times C_L$. From this, it can be observed that the devices in the series slowdown the process. To have a uniform delay and reduced performance, the devices must be made wider. For sizing of the gates, we should pick the worst-case conditions. By following the above procedure, the sizing of the two-input NAND gates is done, as shown in Figure 4.15, where W_p and W_n are the minimum widths of PMOS and NMOS, respectively.

Example 4.2:

Size the gate shown in Figure 4.16(a) to have the minimum delay as the CMOS inverter. Assume minimum widths of PMOS and NMOS are W_p and W_n, respectively.

ANSWER:

In a PUN, three PMOS transistors will be in series in the worst case, so the minimum width of PMOS will be multiplied by 3.
  PMOS width = $3 \times 2\ W_p$
In a PDN, two NMOS transistors are in series, so the minimum width will be multiplied by 2. Finally, the sizing of all transistors is shown in Figure 4.16(b).
  NMOS width = $2 \times W_n$

4.4.2   RATIOED LOGIC

Ratioed logic is used to reduce the number of transistors required to implement any given logic function. In complementary CMOS, it consists of two types of network, the PUN and the PDN, as shown in Figure 4.17. The PUN provides a conditional path between V_{DD} and the output when the PDN is turned off. In ratioed logic, PUN is

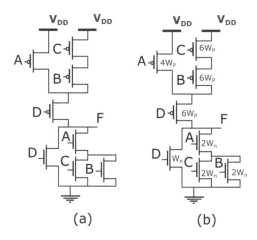

FIGURE 4.16   (a) Static CMOS complex gate. (b) Sizing of CMOS complex gate [6].

replaced by an unconditional load device that helps to reach the output at a nominal high output voltage (V_{OH}). On the other hand, PDN gives a conditional path between output and ground when PUN is turned off. This circuit is known as the pseudo-NMOS circuit, which is an example of the ratioed logic circuit.

In the pseudo-NMOS ratioed logic circuit, the number of transistors required gets reduced. In complementary CMOS, N-input logic requires 2N transistors, but in a pseudo-NMOS logic circuit, the number of transistors required is N + 1. In ratioed logic, the circuit propagation delay and the power dissipation are affected by the size of the transistor, but in complementary CMOS, they are not affected by the size of the transistor. In ratioed logic, it depends upon the ratio of the size of NMOS and PMOS that is why it is known as ratioed logic. The main disadvantage

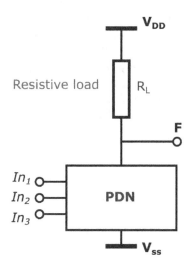

FIGURE 4.17   CMOS-based ratioed logic.

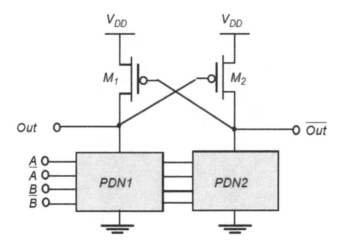

FIGURE 4.18 Topology of DCVSL [2].

of the pseudo-NMOS logic is the high static power consumption due to the direct path that exists from V_{DD} to GND. We can design possible alternative topologies for ratioed logics that can completely eliminate static currents and provide lower power consumption [7–9].

4.4.2.1 Differential Cascade Voltage Switch Logic (DCVSL)

Figure 4.18 shows the basic topology of DCVSL logic [2] that uses both PDN1 and PDN2 mutually exclusive to each other. That means if PDN1 operates, PDN2 switches OFF and vice versa. As a result, single gate implements both logic and its inverse simultaneously. If PDN1 is ON, the output value discharges to GND. Consequently, transistor M_2 switches ON and makes output charges to V_{DD}. As PDN1 and PDN2 are mutually exclusive, in this phase, PDN2 is switched OFF reducing the static power consumption of the resultant gate. This circuit reduces the static power consumption by eliminating the direct path between V_{DD} and GND. However, the logic is still ratioed and strongly depends on the sizing of PMOS devices. Following a similar design, Figure 4.19 shows the AND/NAND gate that reduces the static power consumption.

4.4.3 Pass-Transistor Logic

Pass transistor logic is a popular alternative to complementary CMOS that aims to reduce the number of transistors required to implement logic by allowing the primary inputs to drive gate terminals as well as source/drain terminals [10]. This is different from logic families that we have studied so far, which only allows primary inputs to drive the gate terminals of metal oxide semiconductor field-effect

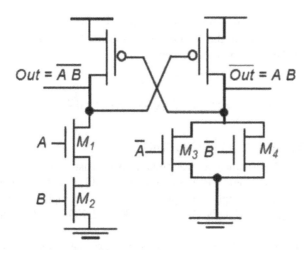

FIGURE 4.19   DCVSL AND/NAND gate.

transistors (MOSFETs). Figure 4.20 shows an implementation of the AND function constructed that way, using only NMOS transistors. In this gate, if input B is high, the top transistor is turned on and copies input A to the output, When B is low, the bottom pass transistor is turned on and 0 is passed.

The switch driven by "B" seems to be redundant at the first glance, but its presence is essential. Figure 4.20 ensures that the gate is static, that is, a low-impedance path exists to the supply rails under all circumstances or when B is low. The advantage of this approach is that fewer transistors are required to implement a given function. In Figure 4.20, AND gate is implemented that requires 4 transistors (including the inverter required to invert B), while a complementary CMOS implementation would require 6 transistors. The reduced number of devices leads to lower capacitance. But an NMOS device is effective at passing a 0, but is poor at pulling a node to V_{DD}. The output can only charge up to $V_{DD}-V_{Tn}$. Also, the situation is worsened by the fact that the devices experience the body effect, as there exists a significant source-to-body voltage when pulling high. Let's take the case

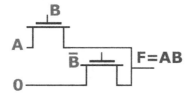

FIGURE 4.20   Pass transistor AND implementation [10].

when the pass transistor is charging up a node with the gate and drain terminals set at V_{DD}. Let the source of the NMOS pass transistor be x. Node x will charge up to $V_{DD}-V_{Tn}$ (V_x). The V_x can be expressed as

$$V_x = V_{DD} - \left(V_{tno} + \gamma \left(\left(\sqrt{|2\varphi_f| + V_x} \right) - \sqrt{|2\varphi_f|} \right) \right) \tag{4.18}$$

4.4.3.1 Differential Pass Transistor Logic

For an improved performance, a differential pass transistor logic (DPL) or complementary pass-transistor logic(CPL), is commonly used [7]. Its basic idea is to accept true and complementary inputs and produce true and complementary outputs. A number of CPL gates (AND/NAND, OR/NOR, and XOR/XNOR) are shown in Figure 4.21.

Since the circuits are differential, complementary data inputs and outputs are always available. For generating the differential signals, extra circuitry is required, but CPL has the advantage that some complex gates such as XORs and adders can be realized efficiently with a small number of transistors. Also, the availability of both polarities of each signal eliminates the need of extra inverters, as it is often the case in static CMOS or pseudo-NMOS. The CPL belongs to the group of static gates because the output nodes are always connected to either V_{DD} or GND through a low resistance path. This is advantageous for the noise resilience.

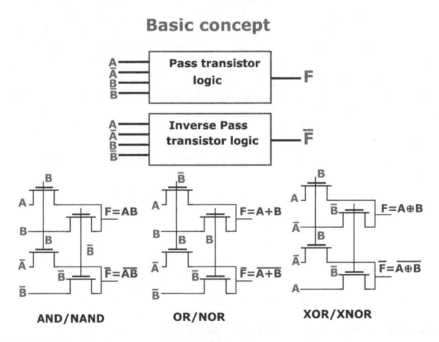

FIGURE 4.21   Complementary pass transistor logic.

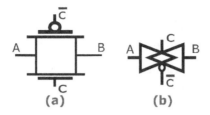

FIGURE 4.22   Transmission gate (a) Circuit. (b) Symbolic representation.

4.4.3.2   Transmission Gate Logic

In the latest submicron CMOS technologies, the problem of the voltage drop on the IC supply rails has become significant. The power consumption has remained almost the same, or is even higher than it used to be, because integrated circuits have more gates and run at higher frequencies. This means, for the same power, the currents flowing in the power supply are comparatively higher. Transmission gate logic is used to deal with the voltage-drop problem [8]. It is based on the complementary properties of NMOS and PMOS transistors. NMOS devices are used to pass a strong 0 but a weak 1, while PMOS transistors pass a strong 1 but a weak 0. Generally, NMOS acts as a pull-down resistor and PMOS as a pull-up. In Figure 4.22(a), the transmission gate acts as a bidirectional switch controlled by the gate signal C. When C = 1, both NMOS and PMOS are on, allowing the signal to pass through the gate. In short, A = B if C = 1. When C = 0, both transistors are in cutoff, an open circuit is created between the nodes A and B. Figure 4.22(b) shows a commonly used transmission-gate symbol.

Consider the case when charging node B to V_{DD} for the transmission gate circuit in Figure 4.23(a). Node A is driven to V_{DD} and the transmission gate is enabled (C = 1 and C = 0). If only the NMOS pass device is present, node B charges up to $V_{DD}-V_{Tn}$, at the point when the NMOS device turns off. Since the PMOS device is present and turned on ($V_{GSp} = -V_{DD}$), charging continues up to V_{DD}. Figure 4.23(b) shows the opposite case that is discharging node B to 0. B is initially at V_{DD} when node A is 0. The PMOS transistor itself can only pull down B to V_{Tp} at which it turns off. The parallel NMOS device stays turned on (since $V_{gs} = V_{DD}$) and pulls node B to the GND. As the transmission gate requires two transistors and more control signals, it enables rail-to-rail swing.

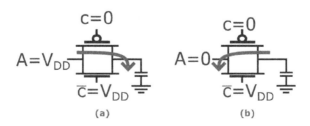

FIGURE 4.23   Transmission gate operation (a) charging node B (b) discharging node B.

TABLE 4.2

Truth Table of 2 × 1 Multiplexer

Selection Line (S)	Output (Y)
Logic 0	A
Logic 1	B

Example 4.3:

Design a transmission gate-based 2 × 1 multiplexer? Draw the layout for it?

ANSWER:

First write the truth table of multiplexer (Table 4.2) where A, B are the inputs and S is the selection line.

From the truth table, the Boolean equation can be expressed as

$$Y = A\overline{S} + BS$$

The transmission gate-based design is shown in Figure 4.24.

4.5 CMOS SEQUENTIAL DESIGN

This section briefly describes the designing of sequential logics with a suitable example.

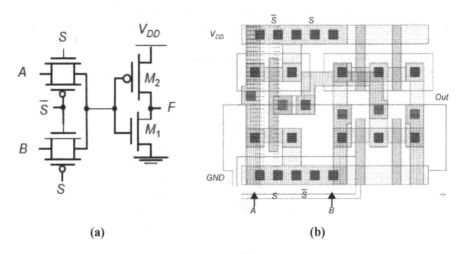

(a)   (b)

FIGURE 4.24   (a) Transmission gate-based MUX design (b) layout.

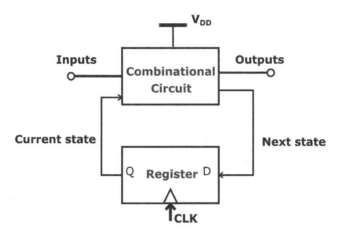

FIGURE 4.25 Block diagram of sequential circuits.

4.5.1 INTRODUCTION

Combinational circuits have a certain logic and process the input signals accordingly to produce the output. On the other hand, for sequential circuits, the output not only depends on the present input but also on the previous output. That means the sequential circuit remembers some past history in its memory. The block diagram of the sequential circuit is shown in Figure 4.25, wherein it consists of combinational circuit and memory. The output of this circuit depends on the present input and the state of the memory element present in it. In general, these memory elements work with respect to the clock signal.

4.5.2 METRICS FOR CMOS SEQUENTIAL DESIGN

Majorly, there are three important timing metrics for memory, here the memory is considered a register since it is the smallest memory element. These metrics are illustrated using a diagram, as shown in Figure 4.26 and the timing components are described below:

Setup time: The minimum time required for the input to settle before the clock transition is the setup time (t_{su}) of the register.

Hold time: The minimum time required to remain the input data after the clock transition is known as the hold time (t_{hold}) of the register.

Propagation delay: The time required for the register to propagate the input to the output with the clock transition is the propagation delay (t_d) of the register. The condition for the proper operation of the register is expressed with the minimum clock period (T) is related in equation (4.19).

$$T \geq T_d + T_{su} + T_{hold} \tag{4.19}$$

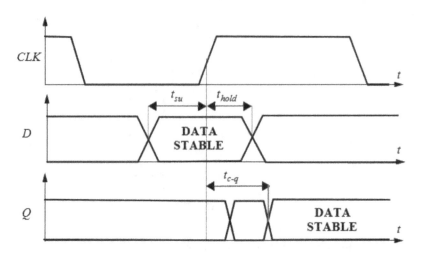

FIGURE 4.26   Pictorial representation of setup, hold, and propagation delay of a register.

4.6   STATIC LATCHES AND REGISTERS

4.6.1   THE BISTABILITY PRINCIPLE

The static memories utilize bistable circuits, that is, the circuits have two stable states (state 0 and state 1). The basic circuit is shown in Figure 4.27(a), wherein two inverters are connected in a cascaded fashion. The VTC plot of the first inverter and the plot of the second inverter are shown in Figure 4.27(b). When we combine both the inverters as shown in Figure 4.27(b), the combined VTC will have two stable states and one metastable state.

Hence, the cross-coupled inverter pair has two stable states and serves as a memory element which can store either 0 or 1. To change the state, we have to force the

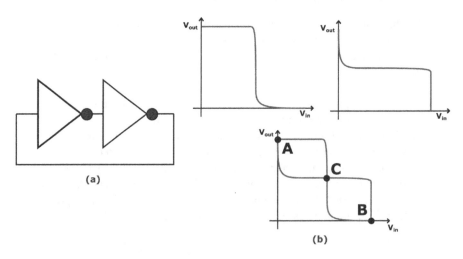

FIGURE 4.27   (a) CMOS cascaded inverter. (b) VTC characteristics.

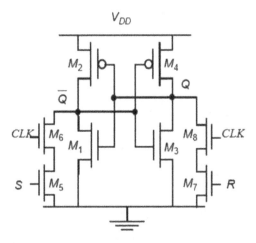

FIGURE 4.28   CMOS-based SR flip-flop.

triggered pulse at the input of one inverter, which changes the state of the memory and these bistable circuits are also called flip-flops.

4.6.2   SR FLIP-FLOPS

Figure 4.28 shows the SR flip-flop [9] that consists of a cross-coupled inverter pair and extra transistors to change the state and to apply clock signal. When clock (CLK = 1), with signal S = 1 and R = 0, the transistors M_5, M_6 will be switched ON that makes the node $\bar{Q}$ to logic 0. With the functionality of the back-to-back inverter pair, the output Q is set to logic 1. Similarly, when $R = 1$ and $S = 0$, M_7 and M_8 switch ON that makes the node Q to logic 1, $\bar{Q}$ to logic 0.

4.6.3   D-LATCHES AND FLIP-FLOPS

There are many ways available to design D-latch, in which a simple method is the transmission gate-based D-latch. Figure 4.29 shows the transmission gate-based

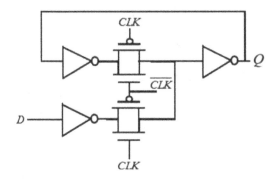

FIGURE 4.29   Transmission gate-based simple D-latch.

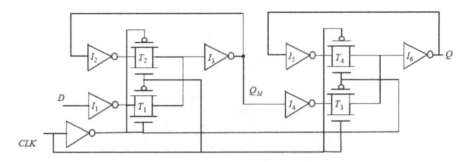

FIGURE 4.30 Transmission gate-based D-flip flop.

simple positive D-latch design. When the CLK is high, the latch propagates the input to output. When the CLK is low, the latch retains the output with back-to-back connection of inverters. In this phase, the latch doesn't have any connection with the input. So, the previous output logic propagates through the back-to-back inverter pair until the next positive CLK.

4.6.4 MASTER-SLAVE FLIP-FLOP

Figure 4.30 shows the transmission gate-based master-slave D flip-flop where positive and negative D-latches are combined. At a low CLK signal ($\overline{CLK} = 1$), the T_1 transmission gate switches ON and T_2 switches OFF. As a result, the input D propagates to node Q_M. During this phase, the transmission gate T_3 switches OFF and T_4 switches ON. Consequently, the cross-coupled inverters (I_5, I_6) hold the previous state of the slave latch. When the CLK signal is high, the master stage enters into hold mode and stores the previous value due to the cross-coupled inverters I_3 and I_4. In this phase, T_3 switches ON and T_4 switches OFF, and Q_M propagates to the output Q.

4.7 SUMMARY

The outline of this chapter can be summarized as follows:

- This chapter began with the CMOS inverter design and presented the characteristics of the CMOS inverter. The characteristics of the CMOS inverter are benchmarked against the NMOS inverter, which proves the robustness of the CMOS inverter.
- The static and dynamic behaviors of a CMOS inverter were observed and the propagation delay and power consumption were modeled. This discussion was extended to combinational circuit design using CMOS.
- Different logic families including static CMOS, ratioed logic, pass transistor, and transmission gate logic were presented.
- Further, CMOS-based sequential circuit design was also introduced. This covered the CMOS-based bistability principle and the utility of the CMOS inverter in the sequential circuit design. Finally, CMOS-based latches and flip-flops were introduced and explained.

4.8 MULTIPLE-CHOICE QUESTIONS

1. Which of the following is not a CMOS inverter characteristic?
   a. Low power consumption
   b. High noise margin
   c. Full logic swing
   d. High power consumption

2. In static CMOS PUN consists of
   a. NMOS only
   b. PMOS only
   c. PMOS and NMOS
   d. None

3. What is the full form of CMOS?
   a. Complementary metal oxide semiconductor
   b. Complex metal oxide semiconductor
   c. Complementary multi oxide semiconductor
   d. Complementary metal oxide ssensor

4. N-input static complementary gate takes
   a. 2N transistors
   b. 2N + 1 transistors
   c. 2N + 2 transistors
   d. 2N − 1 transistors

5. Which of the following is ratioed logic?
   a. Static CMOS
   b. Pass transistor logic
   c. Transmission gate logic
   d. DCVSL logic

6. Pass transistor AND gate uses
   a. 4 transistor
   b. 5 transistors
   c. 6 transistors
   d. 7 transistors

7. Which of the following is the main drawback of pass transistor logic?
   a. High area
   b. High power consumption
   c. Low logic swing
   d. High delay

8. Transmission gate logic uses
   a. PMOS
   b. NMOS
   c. PMOS and NMOS
   d. None of the above.

9. CMOS memory consists of
    a.   Back-to-back inverter
    b.   Single inverter
    c.   Transmission gate
    d.   None

10. N-input pseudo NMOS gate takes
    a.   2N transistors
    b.   N + 1 transistors
    c.   2N + 2 transistors
    d.   2N − 1 transistors

4.9   SHORT-ANSWER QUESTIONS

1. What is the difference between a CMOS and an NMOS inverter?
2. What is DCVSL logic? What is its significance?
3. What is a CMOS transmission gate?
4. Design a pass transistor-based AND gate.
5. Design a static CMOS XOR gate.
6. Define noise margin of a CMOS inverter.
7. What is the setup time of a register?
8. What is bistability and how is it useful?
9. Design a half adder using static CMOS logic.
10. What is the switching threshold of a CMOS inverter?

4.10   LONG-ANSWER QUESTIONS

1. Implement the following expression in a full static CMOS logic fashion using no more than 10 transistors.

$$\bar{Y} = (A.B) + (A.C.E) + (D.E) + (D.C.B)$$

2. Design transmission gate-based full adder circuit. Estimate the number of transistors.
3. Consider the circuit shown in the Figure 4.31 given below
    a.   What is the logic function implemented by the CMOS transistor network? Size the NMOS and PMOS devices so that the output resistance is the same as that of an inverter with an NMOS W/L = 4 and PMOS W/L = 8.
    b.   What are the input patterns that give the worst case tpHL and tpLH. State clearly what the initial input patterns are and what input(s) has to be used to make a transition to achieve the maximum propagation delay.

4. What is ratioed logic? Discuss with an example. State the benefits.
5. Consider the Figure 4.32 shown below
    a.   Write the Boolean equations for outputs F and G.
    b.   What function does this circuit implement?
    c.   What logic family does this circuit belong to?

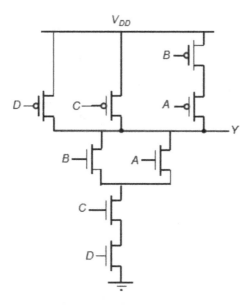

FIGURE 4.31   Static CMOS logic circuit.

6. Design a DCVSL gate which implements the same function. Assume A, B, C, and their complements are available as inputs.
7. Mention the drawbacks of pass transistor logic and discuss the alternatives available.
8. What is the CMOS inverter? Explain the static and dynamic behaviors of the CMOS inverter.

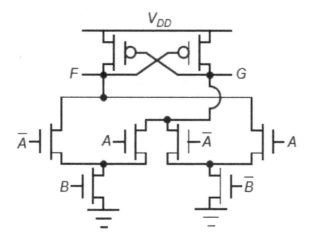

FIGURE 4.32   CMOS logic circuit showing differential output.

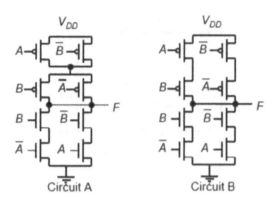

FIGURE 4.33   CMOS circuits for duality test.

9. What is the logic function of circuits A and B in the Figure 4.33 below? Which one is a dual network and which is not? Is the nondual network still a valid static logic gate? Explain. List any advantages of one configuration over the other.

10. Implement the equation $X = ((A + B) (C + D + E) + F) G$ using complementary CMOS. Size the devices so that the output resistance is the same as that of an inverter with an NMOS W/L = 2 and PMOS W/L = 6. Which input pattern(s) would give the worst and best equivalent pull-up or pull-down resistance?

REFERENCES

1. Chandrakasan, A and Brodersen, R. 1995. *Low Power Digital CMOS Design*, Kluwer Academic Publishers, Norwell, MA.
2. Heller, L. et al. 1984. Cascade Voltage Switch Logic: A Differential CMOS Logic Family, *Proc. IEEE ISSCC Conference*, pp. 16–17, February 1984.
3. Chandrakasan, A., Bowhill, W. and Fox, F. (Eds.). 2001. *Design of High-Performance Microprocessor Circuits*, IEEE Press, Piscataway, NJ.
4. Lee, C. M. and Szeto, E. 1986. Zipper CMOS. *IEEE Circuits and Systems Magazine* 2, no. 3: 10–16.
5. Parameswar, A., Hara, H. and Sakurai, T. 1996. A swing restored pass-transistor logic based multiply and accumulate circuit for multimedia applications. *IEEE Journal of Solid State Circuits* SC-31, no. 6: 805–809.
6. Rabaey, J. and Pedram, M. (Eds.). 1996. *Low Power Design Methodologies*, Springer, New York, NY.
7. Shoji, M. 1988. *CMOS Digital Circuit Technology*, Prentice Hall, Upper Saddle River, New Jersey.
8. Weste, N. and Eshragian, K. 1993. *Principles of CMOS VLSI Design: A Systems Perspective*, Addison-Wesley, Boston, MA.
9. Krambeck, R. et al. 1982. High-speed compact circuits with CMOS. *IEEE Journal of Solid State Circuits* SC-17, no. 3: 614–619.
10. Radhakrishnan, D., Whittaker, S. and Maki, G. 1985. Formal design procedures for Pass-transistor switching circuits. *IEEE Journal of Solid State Circuits* SC-20, no. 2: 531–536.

5 Analog Circuit Design

5.1 INTRODUCTION TO ANALOG DESIGN

With the advancement of complementary metal oxide semiconductor integrated circuit (CMOS IC) technology, several analog functions can be digitally implemented [1]. However, a few important applications are still required that can work with only analog signals and those are summarized below:

Operation with natural signals: In reality, the naturally available signals are analog in type. For example, the high-quality microphone generates a sound signal of amplitude varying from few millivolts to hundreds of millivolts. The photo cell in a video camera generates a current signal and a seismograph generates an output voltage ranging from microvolts.

Multilevel signals: To improve the quality of long-distance communication, a system requires multilevel signals rather than the binary signals.

Sensors and actuators: Several sensors dealt with mechanical and electrical signals. These sensors play a critical role in day-to-day life.

5.2 MOS DEVICE FROM ANALOG PERSPECTIVE

In this section, the characteristics of the metal oxide semiconductor field-effect transistor (MOSFET) is analyzed in view of analog circuit design. Further, this section also explains second-order effects of MOSFET and small signal model of MOSFET.

5.2.1 I/V Characteristics

With zero voltage applied at the gate to source (V_{GS}) terminal, no current flows between source and drain with the application of positive drain to source voltage (V_{DS}). With this biasing, the transistor is in the cut-off region and hence zero current flows, as shown in Figure 5.1.

When $V_{GS} = 0$; $I_D = 0$; Cut-off region.

With positive V_{GS} and V_{DS} applied, the gate pushes the positive charge carriers down the body and the full of negative voltage will be formed between the drain and the source. The induced n-region forms a channel with an application of positive V_{DS}, as shown in Figure 5.2 and charge carriers flow between the source and the drain. The minimum value of V_{GS} required to form a channel is called the threshold voltage (V_t).

With the small value of V_{DS}, current varies linearly and increases with an increase in drain voltage. This region is called the triode and in this region, the

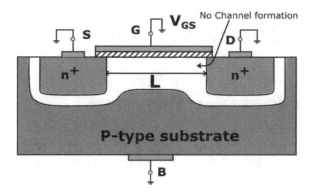

FIGURE 5.1   *N*-channel MOSFET with zero V_{DS}.

circuit works as a resistor. The drain current of MOSFET in the triode region is modeled as below:

When $V_{GS} > V_t$, $V_{DS} < V_{GS} - V_t$; Triode region

$$I_D = \mu_n C_{ox} \frac{W}{L} \left[(V_{GS} - V_t)V_{DS} - \frac{1}{2} V_{DS}^2 \right] \qquad (5.1)$$

With an increased V_{DS}, the channel obtains the tapered shape, as shown in Figure 5.3, and the channel depth at the drain side reduces to zero. This effect is called pinch-off, and current in this region becomes constant. The drain current thus saturates at this value; therefore, this region is named as the saturation region of MOSFET. The drain current of MOSFET in the saturation region is modeled as below:

When $V_{GS} > V_t$, $V_{DS} > V_{GS} - V_t$; Saturation region

$$I_D = \frac{1}{2} \mu_n C_{ox} \frac{W}{L} \left[(V_{GS} - V_t)^2 \right] \qquad (5.2)$$

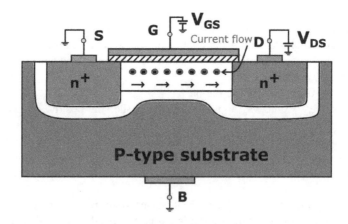

FIGURE 5.2   *N*-channel MOSFET with small V_{DS} value.

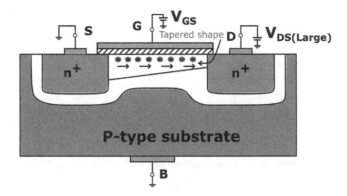

FIGURE 5.3 *N*-channel MOSFET with large V_{DS} value.

The transfer and output characteristics of an n-channel metal oxide semiconductor (NMOS) transistor are shown in Figures 5.4(a) and 5.4(b), respectively. Transfer characteristics are drawn between input voltage (V_{GS}) and drain current; it shows that an NMOS produces the drain current when V_{GS} is greater than the threshold voltage. The output characteristics are drawn between output voltage (V_{DS}) and drain current, and the region of operations is indicated. As discussed in the operation of MOSFET, at a small value of V_{DS}, MOSFET operates in the triode region; when V_{DS} increases, it enters into the saturation region.

5.2.2 Second-Order Effects

In this section, the second-order effects that occur in MOSFET are discussed. These effects are analyzed in order to consider them for MOSFET-based circuit design.

5.2.2.1 Body Effect

In general, we consider MOSFET where source and body terminals are tied together. If the bulk and source terminal voltages of a transistor differ from

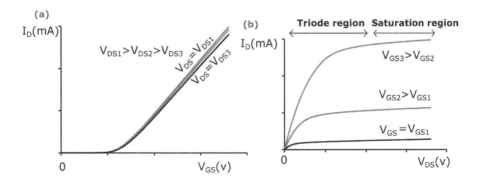

FIGURE 5.4 *N*-channel MOSFET: (a) Transfer and (b) Output characteristics.

each other, body effect comes into the picture. To understand this, when $V_D = V_S = 0$, the V_B becomes more negative that increases the charge density in the channel. Consequently, depletion region becomes wider and the threshold voltage increases. It can be shown that the threshold voltage is expressed in terms of source-to-body voltage as

$$V_{th} = V_{th0} + \gamma \left(\sqrt{|2\phi_F + V_{SB}|} - \sqrt{|2\phi_F|} \right) \tag{5.3}$$

where V_{th0} is the threshold voltage at zero V_{SB}, $\gamma = \frac{\sqrt{2qN_{sub}\epsilon_{si}}}{C_{ox}}$ denotes body effect coefficient and V_{SB} is the source to body voltage.

5.2.2.2 Channel-Length Modulation

In the saturation region of MOSFET after occurrence of pinch-off, if the potential difference between the drain and the source increases, the channel gradually decreases. In other words, the channel length is the function of the drain to the source voltage. Due to the channel length modulation, the drain current can be written as follows:

$$I_D = \frac{1}{2}\mu_n C_{ox} \left(\frac{W}{L} \right)(V_{GS} - V_t)^2 \left(1 + \lambda V_{DS} \right) \tag{5.4}$$

where λ is the channel-length modulation coefficient.

5.2.2.3 Subthreshold Conduction

In general, it is assumed that MOSFET turns OFF as V_{GS} drops below V_{th}. However, in reality, MOSFET exhibits weak inversion where the drain current still exists and some current flows from drain to source. The drain current in this region can be written as

$$I_D = I_o \exp^{\frac{V_{GS}}{\zeta V_T}} \tag{5.5}$$

where $\zeta > 1$ is a non-ideality factor and $V_T = kT/q$.

5.2.3 MOS Small Signal Model

Small signal model of MOSFET is useful in deriving mathematical results with simplified calculation. Figure 5.5 shows the small signal model of MOSFET. As MOSFET has an insulated gate, the gate-to-source impedance is very high. It can be observed that the small signal impedance between the gate and source is modeled as open circuit. As the drain current is a function of the gate-to-source voltage, a voltage-dependent current source ($g_m V_{GS}$) is incorporated in small signal model. In order to cover channel length modulation effect, a linear resistor is added between the source and the drain.

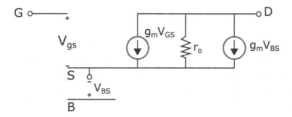

FIGURE 5.5 Small signal model of MOSFET.

The resistance r_o is given by

$$r_o = \frac{\partial V_{DS}}{\partial I_D}$$

$$= \frac{1}{\frac{\partial I_D}{\partial V_{DS}}}$$

$$= \frac{1}{\frac{1}{2}\mu_n C_{ox} \frac{W}{L}(V_{GS} - V_t)^2 .\lambda} \tag{5.6}$$

$$\approx \frac{1}{\lambda I_D}$$

With all terminals at constant voltage, drain current is a function of the body voltage. This effect is modeled as voltage-dependent current source $(g_{mb}V_{GS})$ connected between the drain and the source.

5.3 SINGLE-STAGE AMPLIFIER

This section discusses the low frequency behavior of single-stage CMOS amplifiers [1]. In each case, we start with small signal model and derive the gain of single-stage amplifiers.

5.3.1 COMMON SOURCE

Common source stage is designed by using NMOS FET with a resistive load, as shown in Figure 5.6(a). V_{in} is the input provided to gate terminal of NMOS, and V_{out} is the output of the amplifier collected from drain of NMOS. The small signal

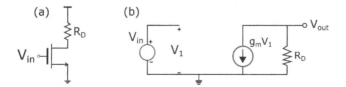

FIGURE 5.6 MOSFET-based (a) common source amplifier and (b) its small signal model.

equivalent of this single-stage circuit is shown in Figure 5.6(b). If the input voltage is below V_{th}, the NMOS transistor switches OFF and the output is V_{DD}. As input voltage increases than V_{th}, the NMOS transistor switches ON. With a further increase in input voltage, output decreases and the transistor enters into the saturation region. In the saturation region of a transistor, the common source stage works as an amplifier. From the small signal equivalent, the gain of the common source stage is derived as follows:

$$V_{out} = -(g_m V_1) R_D$$

$$V_1 = V_{in}$$

From the above both equations, we have

$$V_{out} = -(g_m V_{in}) R_D$$

$$\frac{V_{out}}{V_{in}} = -g_m R_D \qquad (5.7)$$

The negative sign in equation (5.7) indicates the 180° phase shift of the amplifier.

5.3.2 COMMON GATE

In common-gate stage, the input signal is applied to the source of the NMOS and the output is collected from the drain of the NMOS, as shown in Figure 5.7(a). The gate terminal is connected to the dc voltage to establish the proper biasing required. The small signal equivalent of the common gate stage is shown in Figure 5.7(b). The gain of common stage is derived as follows: From signal equivalent, we have

$$-\frac{V_{out}}{R_D} = g_{mb} V_{bs} + \frac{V_{out} - V_{in}}{r_o} + g_m V_1 \qquad (5.8)$$

Since

$$V_{bs} = V_1, \; V_1 = -V_{in} \qquad (5.9)$$

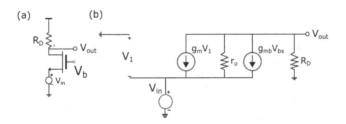

FIGURE 5.7   MOSFET-based (a) common-gate amplifier and (b) its small signal model.

From equations (5.8) and (5.9), we have

$$-\frac{V_{out}}{R_D} = -g_{mb}V_{bs} + \frac{V_{out}-V_{in}}{r_o} - g_m V_{in}$$

$$\left(g_m + \frac{1}{r_o} + g_{mb}\right)V_{in} = \left(\frac{1}{R_D}+\frac{1}{r_O}\right)V_{out} \qquad (5.10)$$

$$\frac{V_{out}}{V_{in}} = \frac{(g_m+g_{mb}+1)R_D}{R_D+r_o}$$

5.3.3 SOURCE FOLLOWER

Source follower stage is also called the common drain stage. Generally, the common drain stage is used as buffer and its gain is approximately equal to unity. Figure 5.8(a) shows the common drain stage, where the input is applied to the gate terminal of NMOS and the output is collected from the source terminal of NMOS. The small signal equivalent of the common drain stage is shown in Figure 5.8(b). The gain of the common drain stage is derived as follows:
From small signal equivalent, we have

$$\frac{V_{out}}{R_s} = g_m V_1 + g_{mb}V_{bs} \qquad (5.11)$$

$$V_{in} - V_1 - V_{out} = 0 \qquad (5.12)$$

Substitute V_1 from equation (5.12) in equation (5.11)

$$\frac{V_{out}}{R_S} = g_m(V_{in} - V_{out}) - g_{mb}V_{out}$$

$$V_{out}\left(\frac{1}{R_S} + g_m + g_{mb}\right) = g_m V_{in} \qquad (5.13)$$

$$\frac{V_{out}}{V_{in}} = \frac{g_m R_s}{1+R_s(g_m+g_{mb})}$$

The obtained gain from equation (5.13) is approximately equal to the unity, and hence the source follower is used as the buffer design.

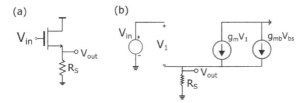

FIGURE 5.8 MOSFET-based (a) common drain amplifier and (b) its small signal model.

5.4   CURRENT MIRRORS

This section introduces the concept of current mirror and its application in analog circuit design. Further, the different current mirror circuits are also analyzed.

5.4.1   INTRODUCTION

The current source plays a vital role in analog design. Current sources in combination with current mirrors can perform useful functions in analog design [2]. Let us consider a MOSFET circuit shown in Figure 5.9 that can act as a current source. Assuming the NMOS transistor is in saturation, we can write the drain current as below:

$$I_{out} = \frac{1}{2}\mu_n C_{ox}\left(\frac{W}{L}\right)\left(\frac{R_2}{R_1 + R_2}V_{DD} - V_{TH}\right)^2 \tag{5.14}$$

It can be observed that the output current depends on supply, process, and temperature. The overdrive voltage depends on both the supply voltage and V_{th}. Threshold voltage varies from wafer to wafer, μ_n and V_{th} also exhibit temperature dependencies. When overdrive voltage becomes low, the current density becomes higher. These variations still exist even if there is no change in gate-to-source voltage. For this reason, we must need another method for biasing of metal oxide semiconductors (MOS) as current source.

The design of current sources in analog circuits follows a method of copying, assuming that precise reference current source is available. A relatively complex circuit is used to generate a stable reference current (I_{REF}), which is then copied to many current sources in the system. Let us consider a copying circuit, as shown in Figure 5.10 and analyze how current can be copied from a reference. For a MOSFET,

$$\begin{aligned} \text{If } I_D &= f\ (V_{GS}), \\ \text{then } V_{GS} &= f^1(I_D) \end{aligned} \tag{5.15}$$

This implies that if this gate-to-source voltage is applied to the targeted transistor, $I_{out} = I_{REF}$. From the other sense, the two identical MOS transistors operate in the saturation region with equal gate-to-source voltages possessing equal currents.

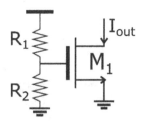

FIGURE 5.9   Current source by using resistive divider.

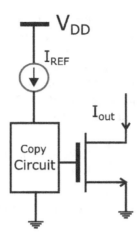

FIGURE 5.10 Concept of copying current.

5.4.2 Basic Current Mirror

The structure of current mirror consisting of M_1 and M_2 transistors is shown in Figure 5.11. In general, the devices are never identical. Neglecting channel length modulation, we can write drain currents as

$$I_{REF} = \frac{1}{2}\mu_n C_{ox} \left(\frac{W}{L}\right)_{M1} (V_{GSM1} - V_{TH})^2 \tag{5.16}$$

$$I_{REF} = \frac{1}{2}\mu_n C_{ox} \left(\frac{W}{L}\right)_{M2} (V_{GSM2} - V_{TH})^2 \tag{5.17}$$

From equations (5.16) and (5.17), we have

$$I_{out} = \frac{\left(\frac{w}{L}\right)_{M2}}{\left(\frac{w}{L}\right)_{M1}} I_{REF} \tag{5.18}$$

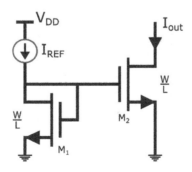

FIGURE 5.11 Simple CMOS current mirror.

From equation (5.18), it is shown that the circuit allows copying of current with no dependence on the process and temperature. In other words, the magnitudes of I_{out} and I_{REF} will be equal if the size of the devices used is identical.

So far, the analysis of the current mirror is performed by neglecting the channel length modulation. Neglecting this effect leads to significant error in copying currents, especially with lower channel length MOSFETs. For current mirror shown in Figure 5.11, the drain current can be written as

$$I_{REF} = \frac{1}{2}\mu_n C_{ox} \left(\frac{W}{L}\right)_{M1} (V_{GSM1} - V_{TH})^2 (1 + \lambda V_{DSM1}) \tag{5.19}$$

$$I_{REF} = \frac{1}{2}\mu_n C_{ox} \left(\frac{W}{L}\right)_{M2} (V_{GSM2} - V_{TH})^2 (1 + \lambda V_{DSM2}) \tag{5.20}$$

From the above equations

$$\frac{I_{out}}{I_{REF}} = \frac{\left(\frac{W}{L}\right)_{M2} (1 + \lambda V_{DSM2})}{\left(\frac{W}{L}\right)_{M1} (1 + \lambda V_{DSM1})} \tag{5.21}$$

when $V_{DSM1} = V_{GSM1} = V_{GSM2}$, V_{DS2} shall not equal to V_{GSM2} because the M_2 transistor is integrated with another circuitry.

5.4.3 CASCODE CURRENT MIRROR

In order to suppress the channel length modulation effect, a cascode current mirror is used, as shown in Figure 5.12. Cascode current mirror consists of three transistors: M_1, M_2, and M_3. The transistor M_3 shields the bottom transistor from variations at P. From Figure 5.12, if V_b is chosen such that $V_Y = V_X$, then I_{out} closely tracks I_{REF}.

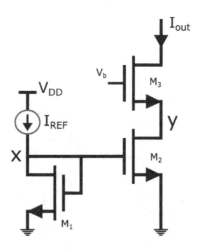

FIGURE 5.12   Cascade current mirror.

5.5   DIFFERENTIAL AMPLIFIERS

Differential operation has become the most dominant choice due to several useful properties. The differential amplifier is the most important circuit for analog and mixed signal applications following this principle.

5.5.1   SINGLE-ENDED AND DIFFERENTIAL OPERATION

A single-ended signal is measured from a fixed point with respect to the ground. A differential signal is defined as the output measured from two equal and opposite fixed-point voltages. These two nodes must exhibit identical impedance; both single and differential signaling concepts are shown in Figure 5.13.

Significance and application of differential signaling:

- The advantage of differential operation over single-ended signal is that it is highly robust toward the environmental noise.
- Due to the capacitive coupling between the two adjacent lines, wherein one high-speed clock line disturbs the single-ended signal. If we distribute the signal and place the clock signal in the middle of two signals, the clock signal disturbs the two signals in equal amount.
- The differential signaling rejects the noise that occurs due to supply voltage.
- Differential signaling provides higher signal swings and higher linearity.

5.5.2   BASIC DIFFERENTIAL PAIR

Basic differential pair is designed using two single-ended signal paths, as shown in Figure 5.14. This differential circuit offers high rejection to supply noise and higher outputs. To avoid common mode disturbance, a current source (I_{SS}) is used. When $V_{in1} = V_{in2}$, the bias current of each transistor is equal to $I_{SS}/2$.

5.5.2.1   Input–Output Characteristics

If V_{in1} is more negative than V_{in2}, M_1 switches off, M_2 switches ON, and the current (I_{SS}) flows through M_2. Due to this, the output $V_{out1} = V_{DD}$ and $V_{out2} = V_{DD}-R_D I_{SS}$. When V_{in1} is closer to V_{in2}, the output V_{out1} reduces. When $V_{in1} = V_{in2}$, the output $V_{out1} = V_{out2} = V_{DD}-R_D I_{SS}/2$. As V_{in1} becomes more positive, M_2 switches OFF, $V_{out2} = V_{DD}$. Figure 5.15 shows the input–output characteristics of basic differential pair. It can be observed that as the difference between the inputs increases, the differential output is saturated between the two values.

FIGURE 5.13   Single-ended and differential measurement.

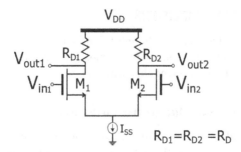

FIGURE 5.14 Basic differential pair.

5.5.3 DIFFERENTIAL PAIR WITH MOS LOAD

The load of a differential pair can be implemented using diode connected or current-source loads. Figure 5.16(a) shows the diode-connected load-based differential pair design, wherein the PMOS transistors are used as diode connected loads. Similarly, Figure 5.16(b) shows the current-source load-based differential pair design, wherein the PMOS transistors are used as current source loads.

The gain of diode connected differential pair is expressed as

$$A_v = g_{mN} \left(g_{mP}^{-1} \parallel r_{ON} \parallel r_{OP} \right)$$
$$= \frac{g_{mN}}{g_{mP}} \tag{5.22}$$

where g_{mN} and g_{mP} denote the transconductance of NMOS and PMOS, respectively. Substituting g_{mN} and g_{mP} in terms of device dimensions, we have

$$A_v = \sqrt{\frac{\mu_n \left(W/L \right)_N}{\mu_p \left(W/L \right)_P}} \tag{5.23}$$

The gain of current source load-based differential pair is expressed as

$$A_v = -g_{mN} \left(r_{ON} \parallel r_{OP} \right) \tag{5.24}$$

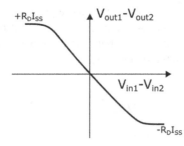

FIGURE 5.15 DC characteristics of differential pair.

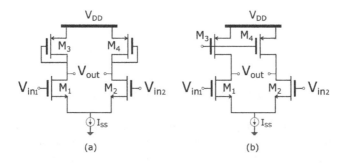

FIGURE 5.16   Differential pair with (a) diode connected load and (b) current source load.

5.6 OPERATIONAL AMPLIFIER

Operational amplifiers (op-amp) are the important subsystems of analog integrated circuits [1]. As their name signifies, op-amp is used to implement different functions ranging from dc bias generation to analog-to-digital converter. This section begins with the fundamentals of op-amp and two-stage design of op-amp. Further, this section also introduces the comparator design.

5.6.1 FUNDAMENTALS AND GENERAL OP-AMP METRICS

Op-amp is a high gain amplifier that is used in variety of applications. Op-amp primarily consists of two inputs, inverting (V_{inv}) and non-inverting input (V_{non}), and an output terminal. The symbol of op-amp is shown in Figure 5.17. When V_{non} is greater than V_{inv}, op-amp produces highest positive voltage as an output. Otherwise, it produces lowest voltage as output.

Gain: The ideal open-loop gain of the op-amp is infinity and practically it should be very high. While maintaining a high open-loop gain, the op-amp bandwidth and voltage swings of the output also need to be taken care of for proper circuit design.

Bandwidth: Working of the op-amp in wide range of frequencies ranging from tens of *Hz* to *MHz* is required to utilize op-amp in different applications.

Offset: It is defined as the voltage correction required by an op-amp to force its output to zero. In op-amp design, several devices contribute to offset generation and the op-amp should have a lower offset voltage.

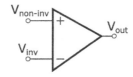

FIGURE 5.17   Block diagram of op-amp.

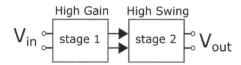

FIGURE 5.18 Block diagram of two-stage configuration of op-amp.

Supply Rejection: Op-amps are sometimes connected to noisy supply voltage. The proper performance of op-amp is required under the influence of noise sources, especially at high frequencies. To obtain this op-amp, always use a differential topology.

5.6.2 TWO-STAGE OP-AMP

In order to avoid the problems raised by a single-ended op-amp design, two-stage op-amp design is introduced. Figure 5.18 shows the block diagram of two-stage configuration of op-amp where first stage provides high gain and second stage provides high swing.

Figure 5.19 shows the simple implementation of two-stage op-amp; each stage can incorporate various amplifier stages, the second stage especially designed as common-source stage to allow maximum output swing. The gain of individual stages is defined as

$$A_{v1} = g_{m1,2}\left(r_{o1,2} \| r_{o3,4}\right) \tag{5.25}$$

$$A_{v2} = g_{m5,6}\left(r_{o5,6} \| r_{o7,8}\right) \tag{5.26}$$

The overall gain can be expressed as

$$A_v = g_{m5,6}\left(r_{o5,6} \| r_{o7,8}\right) \times g_{m1,2}\left(r_{o1,2} \| r_{o3,4}\right) \tag{5.27}$$

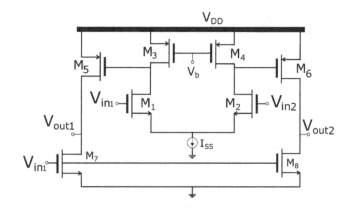

FIGURE 5.19 MOSFET based two-stage configuration of op-amp.

5.7 DIGITAL-TO-ANALOG AND ANALOG-TO-DIGITAL CONVERTERS

This section introduces the concept of both analog-to-digital (ADC) converters and digital-to-analog converters (DAC) [3]. Moreover, the different types of converters have also been explained.

5.7.1 INTRODUCTION

5.7.1.1 DAC: Digital-to-Analog Converter

A DAC takes an n-bit input and produces analog output that is proportional to digital bit sequence. The analog output with digital bits can be expressed as

$$V_{out} = V_{FSR} \left(b_1 2^{-1} + b_2 2^{-2} + \cdots\cdots + b_n 2^{-n} \right) \tag{5.28}$$

where V_{FSR} is the full-scale range voltage, depending on the input pattern, and V_{out} can have 2^n different values ranging from 0 to V_{FSR}.

5.7.1.2 ADC: Analog to Digital Converter

An ADC shows the inverse operation to a DAC; it takes an analog input and produces digital output corresponding to analog voltage. The analog input with digital output bits can be expressed as

$$V_{in} = V_{FSR} \left(b_1 2^{-1} + b_2 2^{-2} + \cdots\cdots + b_n 2^{-n} \right) \tag{5.29}$$

5.7.2 TYPES OF DIGITAL-TO-ANALOG CONVERTERS

Wide varieties of architectures of digital-to-analog converters (DAC) are available. In this section, we will explain the most common architectures of DAC.

5.7.2.1 Weighted Resistor DAC

An n-bit DAC requires n switches, n binary-weighted variables, n-bit summer, and a reference. Figure 5.20 shows the architecture of a weighted resistor n-bit DAC. It uses an op-amp design to sum binary-weighted currents derived from reference voltage via scaling resistors ranging from $2R$ to $2^n R$. The current i_k ($k = 0\ldots.n$) appears on the resistor branch depending on whether the switch is closed or not. The output voltage (V_o) is expressed as

$$V_O = \left(\frac{-R_f}{R} \right) V_{REF} \left(b_1 2^{-1} + b_2 2^{-2} + \cdots\cdots + b_n 2^{-n} \right) \tag{5.30}$$

This simple architecture of a DAC has drawbacks: nonzero resistance of switches and the exponential rise in resistance. The effect of switch resistance disturbs the binary-weighted relation. Moreover, the large resistance in the least significant position makes unrealistic. For example, an 8-bit DAC requires resistances ranging from 2R to 256R.

FIGURE 5.20 Weighted resistor DAC.

5.7.2.2 Weighted Capacitor DAC

MOS-based integrated circuits and microprocessor require on-chip data converter that uses only MOSFETs and capacitors. The DAC architecture based on weighted capacitor is shown in Figure 5.21. This architecture mainly consists of an array of capacitances, and the operation of weighted capacitance DAC alternates between two cycles called reset and sample cycles. During the reset, all the switches shown in architecture are connected to ground, which completely discharges the capacitors. During the sample cycle, SW_0 is opened and all the remaining switches are either connected to the ground or connected to reference voltage (V_{REF}), depending on whether the digital bit is 0 or 1. The result is the redistribution of charge on capacitor and the corresponding output voltage is obtained. The output of the DAC is defined as

$$V_O = V_{REF}\left(b_1 C + b_2 \frac{C}{2} + \cdots\cdots + b_n \frac{C}{2^{n-1}}\right) \qquad (5.31)$$

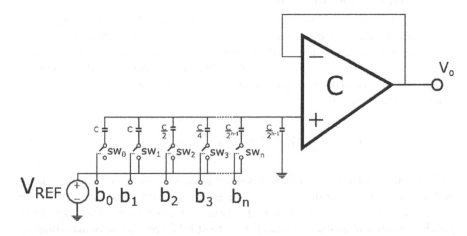

FIGURE 5.21 Weighted capacitor DAC.

As the ratio of the capacitance can be easily controlled to 0.1% accuracies, the weighted capacitor DAC can be designed up to $n \leq 10$. The main drawback of this DAC is exponentially increasing capacitance values.

5.7.3   Types of Analog-to-Digital Converters

This section introduces popular analog-to-digital converters (ADC) that are flash, and successive approximation converters [3]. The principle of operation for ADCs is explained with architectures.

5.7.3.1   Flash Converters

The flash converter consists of a resistor string to create 2^{n-1} reference levels and high-speed latched comparators to simultaneously compare V_{in} against each level, as shown in Figure 5.22. To handle analog signals, two resistors, 1.5R and 0.5R, are required. The comparators whose reference value is below V_{in} will produce the output as logic 1, and the remaining ones as logic 0. The resultant code is referred as the thermometer code and is converted to the desired output code using a priority encoder. Input sampling and latching takes place in the first phase of the clock

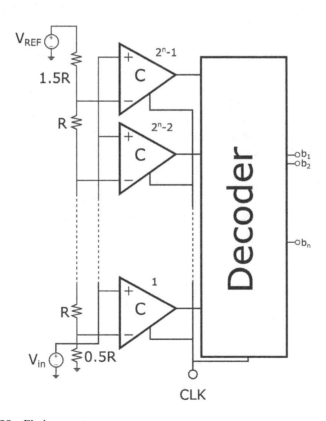

FIGURE 5.22   Flash converter.

period and decoding occurs during the second phase of the clock period. As entire conversion takes only one clock cycle, this ADC is the fastest one. Flash ADC is used for high-speed applications such as video processing and radar communications wherein the conversion rates are higher [3, 4].

The high-speed advantage of flash ADC is the limited utility of $2^n - 1$ comparators. For example, an 8-bit converter requires 255 comparators. The exponential increase in the area and power consumption make flash converter impractical for $n > 10$.

5.7.3.2 Successive-Approximation ADC (SA ADC)

Successive-approximation converter ADCs are highly suitable for low power applications. Due to their scalability, they are highly suitable for low channel lengths. Figure 5.23 shows the basic block diagram of SA ADC that mainly consists of sample and hold circuit, DAC converter, comparator, and SA logic. The N-bit SA ADC digitalization process needs N clock cycles and is expressed as

Latency/conversion time $= N.T_s$.

If data rate is F_s, internal circuit operates at $N \times F_s$. Internal circuit should operate at very high speed, so the design of DAC and comparator ultimately decide the speed of the SAR ADC.

Sample and hold circuit: It is a passive circuit that samples the input data at a clock rate. A sample and hold (S&H) circuit needs to be designed properly because any error occurring in the S&H circuit cannot be recovered further.

DAC and comparator design: A DAC is used to convert the digital output to analog signal to compare it with the analog input of SA ADC. A DAC can be of two types: resistive and capacitive. To compare these analog inputs and the calibrated analog signal, a comparator design is used.

SA algorithm: The algorithm of SA is shown in Figure 5.24. The samples of V_{in} are compared with $V_{D/A}$; then the output of compared is either 1 or 0. If the output is 0, the output of $V_{D/A}$ is subtracted by using $V_{ref}/2^i+1$. Otherwise, the output of $V_{D/A}$ is added by using $V_{ref}/2^i+1$ and comparison takes place. After this, again comparison takes place depending on the output of the DAC.

Working: The working of an SA ADC is divided into two phases: sampling and conversion phases. During the sampling phase, the input V_{in} will be sampled

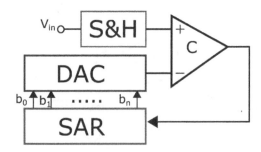

FIGURE 5.23 Basic block diagram of successive approximation ADC.

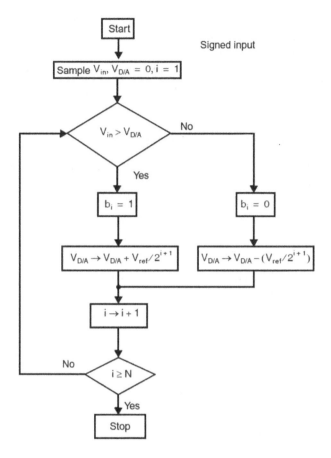

FIGURE 5.24   Successive approximation algorithm flow chart.

through switches S&H circuit. This sampled input will be compared by the comparator design. During the conversion phase, by using SAR algorithm and DAC, the SA produces the digital output related to the analog voltage.

5.8   SUMMARY

This chapter can be summarized as follows:

- In this chapter, we discussed the significance of analog circuit design. First, the MOSFET device characteristics and second-order effects were discussed.
- The single-stage amplifiers like common source, common gate, and common drain were designed and the gain of each stage was derived.
- Further, the current mirror circuits were introduced. The functioning and characteristics of current mirrors were discussed.

- Later, this chapter dealt with the MOS-based differential amplifier and op-amp designs. The two-stage op-amp was designed and the gain of the two-stage op-amp was calculated.
- Finally, the concepts of data converters such as DAC and ADC converters were introduced. Several important converter designs with different architectures were also discussed.

5.9   MULTIPLE-CHOICE QUESTIONS

1. With an increase in source-to-body voltage, the threshold voltage of MOSFET
   a.   Increases
   b.   Decreases
   c.   Remains same
   d.   None

2. Which of the following is useful in successive approximation analog to digital converter?
   a.   DAC
   b.   Sample and hold circuit
   c.   Comparator
   d.   All of the above

3. Gain of MOS-based common source amplifier is
   a. $-g_m R_D$
   b. $g_m{}^2 R_D$
   c. $g_m R_D{}^2$
   d.   None

4. The open-loop gain of the op-amp should be
   a.   Very high
   b.   Very low
   c.   Medium
   d.   None

5. Differential signaling results in
   a.   High gain
   b.   Low noise effect
   c.   Low voltage operation
   d.   None

6. Op-amp can be readily used as
   a.   Comparator
   b.   Oscillator
   c.   Current mirror
   d.   None

7. The drawback of the weighted resistor DAC is
   a. Large resistance
   b. Large offset
   c. Nonlinearity
   d. Both a and c

8. The drawback of a Flash ADC is
   a. Increased comparators
   b. Low speed
   c. High power consumption
   d. None

9. The advantage of Flash ADC is
   a. Increased comparators
   b. High speed
   c. High power consumption
   d. None

10. The drawback of SA ADC is

    a. High conversion time
    b. High power consumption
    c. High nonlinearly
    d. None

5.10   SHORT ANSWER QUESTIONS

1. Why analog circuits are important?
2. What are the MOSFET second-order effects?
3. What are the regions of operations of MOSFET?
4. Draw the small signal equivalent of MOSFET.
5. What is body effect?
6. What is channel length modulation?
7. What is current mirror?
8. What is the concept of differential signaling?
9. What is the concept of analog-to-digital converter?
10. For an n-bit flash converter, what is the number of comparators required?

5.11   LONG ANSWER QUESTIONS

1. Explain the MOSFET-based common source amplifier. Derive the gain.
2. Explain the MOSFET-based common gate amplifier. Derive the gain.
3. Explain the concept of differential pair. Derive its gain.
4. Explain two-stage op-amp. Derive the gain of MOSFET-based op-amp.
5. Explain the weighted resistor DAC and the weighted capacitor DAC.
6. Explain in detail the successive approximation ADC.

REFERENCES

1. Edgar, L. J. 1930. *U.S. Patent No. 1,745,175*. Washington, DC: U.S. Patent and Trademark Office.
2. Kahng, D. 1960. Design of analog CMOS integrated circuits. In *the Design of analog integrated Circuits*, Behzad Razavi.
3. Bondyopadhyay, P. K. 1998. Moore's law governs the silicon revolution. *Proceedings of the IEEE 86*, no. 1: 78–81.
4. Semiconductor Industry Association (SIA). 1999. *International Technology Roadmap for Semiconductors*, 1999 ed., SIA, San Jose, CA.

6 Digital Design Through Verilog HDL

6.1 INTRODUCTION

This section explains the significance and history of Verilog Hardware Description Language (HDL). First, we start with the definition of Verilog HDL and later explain the background and basic concepts of Verilog.

6.1.1 WHAT IS VERILOG HDL?

Verilog HDL is a hardware description language that is utilized in designing of digital systems. The designing may occur at different levels starting from the switch to the top behavioral level. The digital system can be a simple digital gate or a microprocessor design. The programs written in Verilog can be simulated by using several Verilog simulators. These codes are targeted to field programmable gate arrays to implement and test the designs. Due to easy simulation and implementation tools available, the Verilog is significantly used both in industry and academia.

6.1.2 BACKGROUND

In 1983, Gateway Design Automation developed the Verilog HDL as a hardware modeling language for its own simulator. Due to its high popularity and usage, this language later achieved wide attention. In 1995, this Verilog language became standardized according to the Institute of Electrical and Electronics Engineers (IEEE). The complete standard of Verilog HDL is described and the standard is called IEEE Standard 1364-1995 [1].

6.1.3 COMPILER DIRECTIVES

A compiler directive provides certain information that remains through the compilation process until other compiler directives are specified. The syntax of compiler directives starts with a backquote (`) character. The main different compiler directives in Verilog are listed here:

`define, `undef
`timescale
`resetall
`include

`*define, `undef*`

The `define directive is similar to the #define in C programming. This directive defines a macro and is used to substitute text at different places of the programming. In contrast to this, the `undef directive removes the effect of defined macro. The syntax and the usage of macro is shown below:

`define array_size 40

....

reg [array_size-1:0] y;

....

`undef array_size

`timescale

In Verilog HDL, all the delays mentioned need a time unit, and this time unit is mentioned using `timescale compiler directive. This directive is used to specify the unit of time and precision. The syntax of this directive is as follows:

`timescale *time_unit/time_precision*

Example:

`timescale 1 ns/100 ps

This indicates the time unit of 1 ns with the precision of 100 ps

`resetall

This compiler directive resets all reaming directives to their default values. The syntax is as follows:

`resetall

`include

This compiler directive is used to include the content of other files by using their name or the path of the file. The syntax is as follows:

`include "add/work/copy/fulladder.v

Upon execution, this comment replaces the contents of the file fulladder.v.

6.1.4 DATA TYPES

We mainly have two data types, namely net type and variable type [1], the details are as follows:

Net type: This represents a wire connection between the gates and its value depends on the driver of this net. If no gate is connected to the net, its output results in high impedance state.

Variable type: Variable type is used to store the values like register. This variable can be used inside always or initial. The default value of variable is don't care (X).

6.1.5   OPERATORS

There are several different operators available in Verilog as listed below:

Arithmetic operator
Equality operators
Relational operators
Logical operators
Bitwise operators
Conditional operators
Concatenation operators

6.1.5.1   Arithmetic Operator

The different arithmetic operators are listed below:

- +(addition)
- -(Subtraction)
- *(Multiplication)
- /(divide)
- %(modulus)
- **(power)

6.1.5.2   Equality Operators

Different equality operators are listed here:

- ==(logical equality)
- !=(logical inequality)
- ===(case equality)
- !==(case inequality)

If the operands are logically equal, then the comparison results in 0, otherwise 1. In the case of comparisons, x and z are treated as values and compared like numbers. In normal comparison, if x and z are present, then the values returned is x or z.

6.1.5.3   Relational Operators

Several relational operators used in Verilog are listed below:

- $\gg$ (greater than)
- < (less than)
- >= (greater than or equal to)
- <= (less than or equal to)

The relational operator results in 0 if the comparison becomes false, else 1. In case the operands have x and z, then the result is x.

6.1.5.4   Logical Operators

The logical operators of Verilog are summarized below:

- && (logical and)
- || (logical or)
- ! (unary logical negation)

Example:

a=1′b0, b=1′b1
Then: a&&b is 0
   a||b is 1
   If the operands are vectors, the nonzero numbers are considered to be 1. For example, consider two vector operands a= 4′b0010, b=4′b0100
   a&&b is 0 (false)
   a||b is 1 (true)

6.1.5.5   Bitwise Operators

The different bitwise operators are listed below:

- ~ (unary negation)
- & (binary and)
- | (binary or)
- ^ (binary xor)
- ~^ (binary xnor)

Example:

Consider a= 4′b0111, b=4′b0101
   then a&b is 0101
      a|b is 0111

If the operands are unsigned with different sizes, then the small size operand is zero-filled on the most significant bit side. If the operands are signed, the small operand is sign-extended before the operation is performed.

6.1.5.6   Conditional Operator

The syntax of the conditional operator is presented below:

$$Condition? \; expression1: \; expression2$$

If the condition in the above syntax is true, then expression1 will be executed. Otherwise, expresseion2 will be executed. If the condition results in x or z, then the result is a bitwise operation on expression1 and expression2.

Example:

$$Y = a==b? \ a+b: \ a-b;$$

If *a* and *b* are equal, then *a* and *b* will be added. Otherwise, these will be subtracted.

6.1.5.7 Concatenation Operator

Concatenation operates on individual bits and combines them as a string. It is expressed as:

$$\{expression1, \ expression2, \ ....expressionN\}$$

Example:

Consider a= 4'b0111, b=4'b0101
        then y={a, b}; result will be 01110101

6.2 MODULE AND TEST BENCH DEFINITIONS

Verilog consists of a module and a test bench. A module is used to describe the design and a test bench is used to provide the stimulus to test the design described in the module [1]. The module is instantiated in the test bench to show which test bench is associated with a particular module. The detailed syntax and significance of both the module and the test bench designs are described in the following subsections.

6.2.1 MODULE

A module block starts with the keyword 'module' and ends with the keyword 'end-module'. In the module definition, all the input and output terminals of modules are declared. The body of the module consists of a design description that will be completely written in Verilog. The syntax of the module definition is shown below:

```
module module_name (Output1, Output2, ......Outputn, input1, input2,
    ....inputn);
output variable_1;
input variable_1, variable_2,...............variable_n;
begin
    statement_1;

    statement_2;
    ..............
    ...............
    Statement_n;
end
endmodule
```

Important instructions for scripting Verilog module:

- Each module is identified by the unique module name, and two modules should never have the same name.
- The module consists of all input and output declaration with keywords 'input' and 'output', respectively.
- Each statement inside the module ends with a semicolon (;).
- The body of the module or the design description starts within the **begin** and **end** statements.
- The module cannot be nested, that is, one module cannot consist of another module inside it.
- The module ends with the keyword **endmodule**.

Example:

Write a module for full adder design.

```
Module fulladder1 (Sout, Cout, a, b, cin);
Output Sout, Cout;
Input a, b, cin;
begin
    fulladder design description
end
endmodule
```

From the above example, it can be observed that the module name is fulladder1. This module has three inputs, namely a, b, cin and two outputs, namey Sout and Cout. Both the outputs Sout and Cout are declared by using the output keyword. The inputs a, b, and cin are declared by using the input keyword. The design description can be written between begin and end statements that depend on the modeling style and will be explained in the coming sections. Finally, the module ends with the **endmodule** keyword.

6.2.2 Test Bench

A test bench contains a Verilog program that is used to generate test patterns that are used to test the main module. Test bench not only generates the test patterns but also applies those to design. From this, the module performance can be analyzed and tested. The syntax of the test bench is as follows:

```
module test_bench;
reg variable_1, variable_2, .......variable_n;
wire variable_1, variable_2, .......variable_n;
instantiation of module;
test patterns;
endmodule
```

Important instructions for scripting Verilog test bench:

- The test bench doesn't consist of any inputs or outputs as in the module design.
- Each statement inside the test bench ends with the semicolon (;).
- The test bench also starts with the keyword '**module**' and ends with the keyword '**endmodule**'.
- The number of register variables and wires declared in the test bench depends on the number of inputs and outputs of the module.
- The test patterns generated here automatically apply to the module which instantiated inside the test bench.
- There are different methods available to generate the test patterns that will be explained in the upcoming sections.

Example:

Write a test bench for full adder design.

```
module fulladder_test;
reg in1, in2, in3;
wire out1, out2;
fulladder1 g1 (out1, out2, in1, in2, in3);
initial
begin
in1=0;
in2=0;
in3=0;
#50 in1=0;
#50 in2=1;
#50 in3=0;
end
endmodule
```

From the above example, it can be observed that the test bench name is fulladder_test. It consists of three register variables and two wires and contains the instantiation of full adder module. This module instantiation starts with module name, followed by label name (g1) and list of registers and wires. All the test patterns are written inside the initial block (as explained in next sections). The #50 represents the delay of 50 units, and after that delay the value of inputs changed. Thus, different test patterns are generated and the test bench ends with the keyword '**endmodule**'.

6.3   GATE-LEVEL MODELING

This section describes the gate-level modeling of Verilog and related concepts. Finally, an example program is introduced that utilizes the gate-level modeling [2].

6.3.1 BUILT-IN PRIMITIVES

There are different built-in primitives available in Verilog and are listed below:

Single and multiple input gates:
Those are not, and, nand, or, nor, xor, xnor
Tristate gates:
Those are bufif0, bufif1, notif0, notif1
MOS switches:
Those are cmos, nmos, pmos
Bidirectional switches:
Those are tran, tranif0, tranif1

6.3.2 SINGLE AND MULTIPLE INPUT GATES

The available single and multi-input gates in Verilog are listed below in **Table 6.1**. These built-in logic gate primitives will have only one output and multiple inputs and their syntax for declaration is shown below:

multiple-input gate keyword *instance_name (output, Input1, Input2,*
...............Inputn);

Here the instance_name is used to uniquely identify the gate declared. The first terminal always signifies the output and remaining are inputs.

Example:

Declare the two-input NAND gate in Verilog, as shown in Figure 6.1.

Here, to declare NAND gate, we have used nand, a keyword with an instance name *g1*. The first terminal declared as output *Q* is followed by the 2-inputs *A* and *B*. The syntax will be as follows:

nand g1(Q, A, B)

TABLE 6.1
List of Built-in Gates and Significance

S.No	Keyword	Significance
1	not	Signifies inverter
2	and	Signifies and gate
3	nand	Signifies nand gate
4	Or	Signifies or gate
5	nor	Signifies nor gate
6	xor	Signifies xor gate
7	xnor	Signifies xnor gate

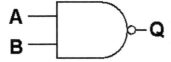

FIGURE 6.1   Symbol of two-input NAND gate.

6.3.3   TRISTATE GATES

Tristate gate operates in three different states namely state 0, state 1, and high imped-ance state (z). Verilog can define tristate gates with the following keywords:

   i. bufif0
  ii. bufif1
 iii. notif0
  iv. notif1

These gates operate with three different states. These gates have one output termi-nal, one input, and a control input. The output of the gate depends on both input and control input. The general syntax of the tristate gate is provided below:

tristate_gate *instance_name* (Output, input, control_input);

The output terminal is declared first, input and control_input follow the output. **Figure 6.2** shows the logic diagrams of different tristate gates, and depending on the

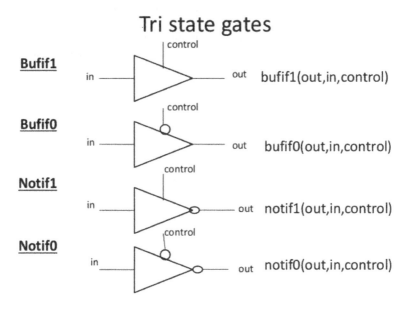

FIGURE 6.2   Logic diagrams of tristate gates.

control input, the output can be driven either to input or to the high impedance state (z). For example, the bufif1 gate output is z if the control input is 0, else input data is transferred to the output. For notif0 tristate gate, the output is z when the control input is 1, else the output is the inversion of input data.

Example:

Declare notif1

$$\text{notif1 } g1(\text{out, in, control});$$

Here, the output of notif1 is inversion of input data if control input is 0, else the output is z.

6.3.4   MOS Switches

Different MOS switches are available in Verilog, as described below. These gates are of unidirectional type and the data flow can be controlled by using the control input.

   i. cmos
  ii. pmos
 iii. nmos

These gates have one input, output, and a control input. The syntax of a MOS switch is shown below:

$$\text{gate\_type } \textit{instance\_name} \text{ (output, input, control\_input);}$$

The first terminal is output and the next is input and finally the control_input is declared at the end. Figure 6.3 shows the symbols of pmos and nmos switches with

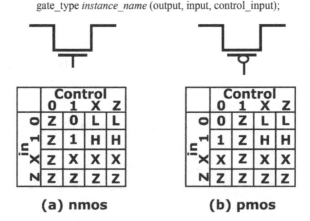

FIGURE 6.3   (a) NMOS, (b) PMOS switches with truth tables.

(b)

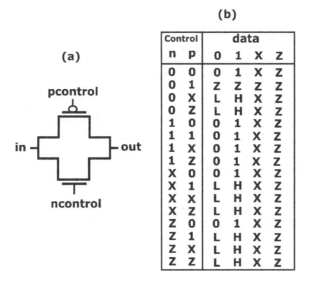

(a)

Control		data			
n	p	0	1	X	Z
0	0	0	1	X	Z
0	1	Z	Z	Z	Z
0	X	L	H	X	Z
0	Z	L	H	X	Z
1	0	0	1	X	Z
1	1	0	1	X	Z
1	X	0	1	X	Z
1	Z	0	1	X	Z
X	0	0	1	X	Z
X	1	L	H	X	Z
X	X	L	H	X	Z
X	Z	L	H	X	Z
Z	0	0	1	X	Z
Z	1	L	H	X	Z
Z	X	L	H	X	Z
Z	Z	L	H	X	Z

FIGURE 6.4   (a) CMOS switch and its (b) Truth table.

their truth tables. If control input is 0, nmos shows high impeadence (z) at the output, else the input data is passed to the output. Whereas with pmos switch, if control input is 0, the input data is passed to the output, else pmos shows high impeadence (z) at the output. Similarly, the cmos switches have one data input, one data output and control input. Figure 6.4 shows the symbol of the cmos switch, and the syntax of the cmos switch is shown below:

cmos *instance_name* (output, input, n_control, p_control);

6.3.5   GATE DELAYS

Gate delay is the propagation delay from the input to the output of any gate. This can be specified with the gate instantiation, and the syntax gate instantiation with gate delay is shown below:

gate_type *gate_delay instance_name* (terminal list);

If the gate delay is not mentioned, it means the propagation delay of the gate is zero. The gate delay consists of three values rise delay, fall delay, and turn_off delay. All the time delay units are specified by using `timescale directive in Verilog HDL. Let us discuss some examples of gate delays.

nor #7 (y, a, b);

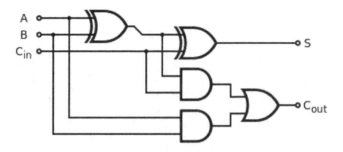

FIGURE 6.5   Gate-level full adder design.

This gate specifies all delays as 7, that is, rise delay and fall delay as 7. Turn_off delay is not applicable to nor gate since the output never goes into high impedance state.

$$nand \ \#(8, 10) \ (y, a, b);$$

In this gate instantiation, the rise delay is specified as 8 and fall delay is specified as 10.

$$bufif1 \ \#(5, 9, 10) \ (y, a, b);$$

In the above tristate gate instantiation, rise delay is 5, fall delay is 8, and turn_off delay is 6.

6.3.6   EXAMPLE

Consider a gate-level design of full adder, as shown in **Figure 6.5**. Gates and intimidated signals are named, as shown in **Figure 6.5**.

A Verilog program of full adder is written by instantiating all the gates in the design, as shown below:

```
module fulladder1 (S, Cout, A, B, Cin);
output S, Cout;
input A, B, Cin;
wire t1, t2, t3;
xor g1 (t1, A, B);
xor g2 (S, Cin, t1);
and g3 (t2, t1, Cin);
and g4 (t3, A, B);
or g5 (Cout, t2, t3);
endmodule
```

6.4   DATAFLOW MODELLING

This section describes the modeling of Verilog program in dataflow style. Here, various aspects of dataflow modeling are discussed with the help of suitable examples.

6.4.1 CONTINUOUS ASSIGNMENT

Dataflow modeling explores continuous assignment statement to assign a value to a net, and its syntax is shown below:

assign LHS_net = RHS_expression;

The above statement uses a keyword assign and the expression on the right-hand side is evaluated and assigns it to the left-hand side net.

Example:

Write continuous assignment statement for the sum output of full adder.

$$sum = a \oplus b \oplus c_{in}$$
assign sum = a^b^cin;

In the above assignment statement, we have used assign and RHS expression as assigned to the LHS net 'sum'. Moreover, the RHS expression utilizes operators to build the expression. These operators differ for different logical and mathematical expressions.

If multiple assignments need to be written in one continuous statement, then the expression looks as provided below. This expression shows both sum and carry of the full adder design.

assign sum= a^b^c, carry= (a&b)|(b&c)|(c&a);

6.4.2 DELAYS

In continuous assignment statement, there is a provision to introduce delay. If no delay is mentioned in the continuous assignment statement, the right-side expression assigns to left side with zero delay. The delay in continuous assignment statement can be introduced as follows:

assign *#delay* target = expression;

The delay provided is the propagation delay between the expression and the target. For example, if the delay value is 7, the expression evaluates and the value is assigned to target after time 7. Similarly, we can write the assignment statement for three delays, as shown below:

assign #(rise, fall, turn-off) target = expression;

Example:

Write different assignment statements by showing different delays.

Statement_1: assign #4 target = expression;
Statement_2: assign # (4, 10) target = expression;
Statement_3: assign # (4, 6, 5) target = expression;

In statement_1, the rise, fall, and turn_off time are observed to be same and it is 4. Whereas in statement_2, the rise time is 4 and the fall time is 10. In the third statement, the rise time is 4, fall time is 6, and turn_off time is 5.

6.4.3   EXAMPLES: A VERILOG PROGRAM FOR FULL ADDER

```
module fulladder1(sout, cout, a, b, cin);
output sout, cout;
input a, b, cin;
assign sout= a^b^cin;
assign cout= (a&b)|(b&cin)|(cin&a);
endmodule
```

6.5   BEHAVIORAL MODELING

In the previous sections, we have seen gate and dataflow modelling of Verilog. Now in this section, we discuss the behavioral-level modeling and related aspects [3]. All the concepts of behavioral-level modeling are explained using example programs.

6.5.1   INITIAL STATEMENT

In modeling behavior of the design, two statements play an important role: initial and always statements. These statements in the Verilog program execute concurrently, which means there is no order for execution of these statements. Initial statement generally executes only once in the program. The syntax of the initial block is as follows:

initial
    begin
procedural_statements;
    end

Execution of procedural statements inside the initial happens only once. The flow execution of procedural statements inside the initial totally depends on the timing delays present in the procedural statements. The initial block is generally used to initialize the variables since it is executed only once. Here is an example of the initial statement for an SR flip-flop.

Example:

Write an initial statement for the SR flip-flop.

In the SR flip-flop, for few cases of input combinations, the output depends on the previous state of the flip-flop. Due to this, we need to initialize the flip-flop state and for that initial block is used as shown below:

initial
begin

```
Q= 0;//the intial state is assumed to be 00 for SR flip-flop and declared
Qb=1;
end
```

Here, inside the initial block, there are two procedural assignment statements that declare the initial state (Q and Qb are the outputs of flip-flop) of flip-flop. These two statements execute sequentially and initial block executes only one for initialization of the flip-flop state.

6.5.2 ALWAYS STATEMENT

Unlike the initial statement, the always statement executes repeatedly. The always statement is very important in behavioral-level modeling and is used for higher-level modeling in Verilog. The syntax of the always statement is as follows:

always @ (*senstivitity list*)
begin
procedural_assignment;
end

Always block starts with the always statement, and it always executes if any variable in the sensitivity list changes. Here, procedural assignment statements execute repeatedly based on the sensitivity list and execute sequentially.

Example:

Write always statement for clock generation with a period of 20 time units

```
always
begin
#10 clk = ~clk;
end
```

This statement executes for infinite time, since there is no condition or sensitivity list available. This always statement produces a clock with a frequency of 20 time units.

Example:

Write always statement with an event control

```
always @ (enable)
begin
a = ~b;
end
```

Statements within begin and end will be executed repeatedly with event occurring on '*enable*'. It means, if any change happened to enable, then only the always block executes.

Example:

Write a Verilog program for D flip-flop

Consider a D flip-flop with an input D, output q, and clock signal clk, and assume that flip-flop is negative edge triggered.

```
module Dflipflop(clk, D, q, qb);
input clk, D;
output reg q, qb;
always @(negedge clk)
begin
D=q;
qb=~q;
end
endmodule
```

In the behavioral level, Verilog modeling all the outputs are declared as registers. Here, as D flip-flop is negative edge triggered, the always block is repeated depending on the negative edge event of the clock signal and it is represented as 'negedge clk'.

6.5.3 PROCEDURAL ASSIGNMENTS

The statements which are written inside the initial or always blocks are called procedural assignment statements. These statements are used to write the behavior of the design inside initial or always. These are mainly categorized into two types:

  i. Blocking procedural assignment
  ii. Non-blocking procedural assignment

6.5.3.1 Blocking Procedural Assignment

The procedural assignment that uses the assignment operator as '=' is called the blocking procedural assignment. The blocking assignment statement executes completely before execution of the next statement.

Example:

Blocking procedural assignment

```
initial
begin
y = #10 0;
y = #20 1;
y= #30 0;
end
```

Here, statement_1 executes and 0 will be assigned to 'y' after 10 time units. After this, the second statement executes and 1 will be assigned to 'y' after 30 time units from 0 time. Finally, statement_3 executes and 'y' is assigned to 0 after 60 time units from 0 time.

6.5.3.2   Nonblocking Procedural Assignment

The procedural assignment that uses the assignment operator as '<=' is called the nonblocking procedural assignment. When the nonblocking assignment statement executes, before completion of assignment, the execution of the next statement can be started based on the delay associated with the statements. It means that the execution will not block at one statement and it continues to the next statement before completion of the present execution.

Example:

Nonblocking procedural assignment

```
initial
begin
y <= #6 0;
y <= #4 1;
y <= #8 1;
end
```

Here, statement_1 starts execution at time 0 and before assignment, the second statement starts execution. After 4 time units, the value of y is assigned to 1, after this at 6 time units, the value of y is assigned to be 0. Finally, the value of y is assigned to be 1 after 8 time units.

6.5.4   CONDITIONAL STATEMENTS

The conditional statements can be used in developing the designs for the behavioral-level modeling. These can be explained in detail in this section with examples.

If statement:
The syntax of 'if' is shown below:

if (*condition_1*)
Procedural assignment_1;
else if (*condition_2*)
Procedural assignment_2;
else
Procedural assignment_3;

First condition_1 is executed, if condition_1 is evaluated to logic '1', then procedural assignment_1 will be executed. If condition_1 is evaluated to logic '0', then the control executes condition_2. If condition_2 is evaluated to 1, then the procedural assignment_2 will be executed, else procedural assignment_3 executes.

Example:

Write if statement for grading a student

```
if (total_marks >75)
begin
grade = A;
a_grade = a_grade+1;
end
else if (total_marks >60)
begin
grade = B;
b_grade = b_grade+1;
end
else
begin
grade = C;
c_grade = c_grade+1;
end
```

First, the condition total_marks >75 is executed, if it is true then the student grade is evaluated to be A. Otherwise, total_marks >60 is executed, if it is true then the student grade is evaluated to be B, else it is evaluated to be C grade.

Case statement:

Case statement is a multiconditional statement and the syntax of the case statement is as follows:

case (*main_expression*)
item_sub_expression: procedural_statement;
item_sub_expression: procedural_statement;
....
item_sub_expression: procedural_statement;
default: *procedural_statement;*
endcase

First, the main expression of the case is evaluated and it is matched with the subexpressions. After matching, the respective procedural statements are executed. In case of no match, the default statement is executed.

Example:

Write a 4 × 1 mux behavior using case

Consider a 4 × 1 mux where a, b, c, and d are the inputs and s, y are the selection line and output, respectively.

```
always @(a or b or c or d or s)
begin
```

```
case(s)
2'b00: y=a;
2'b01: y=b;
2'b10: y=c;
2'b11: y=d;
default: $display("error in selection");
endcase
end
```

Other forms of case statement:

In general, the case statement explained in the previous section, the x and z values are treated as letters. On the other hand, there are other case statements that treat the values of x and z in another manner. Those are casex and casez.

- In casez, the value of z is considered to be don't care.
- In casex, the values of x and z are considered to be don't cares.
- The syntax of both casex and casez are similar to case.

Example:

casez statement for 3-to-8 priority encoder

```
casez (input)
8'b00000001: y= 3'b000;
8'b0000001z: y= 3'b001;
8'b000001zz: y= 3'b010;
8'b00001zzz: y= 3'b011;
8'b0001zzzz: y= 3'b100;
8'b001zzzzz: y= 3'b101;
8'b01zzzzzz: y= 3'b110;
8'b1zzzzzzz: y= 3'b111;
default: y= 3'bzzz;
endcase
```

As we are using casez, all x and z characters are treated as don't cares. If the MSB of the priority encoder input is 1, then the output is y= 111(other bits of input is ignored). If the MSB bit of the input is 0 and its next MSB is 1, then y=110. If no match is found, then y is assigned to don't care.

6.5.5   Loop Statements

The different types of loop statements available are as follows:

  i. For-loop
 ii. While-loop
iii. Forever-loop

6.5.5.1 For-Loop Statement

The syntax of the for-loop is as follows:

for *(initial_assignement; condition; increment assignment)*
begin
procedural statements;
end

A for-loop repeats the execution of procedural statements a number of times depending on the condition of the loop. The increment assignment increases the initialized value and checks the condition with the updated value. As long as the condition is true, the assignment statements specified in the loop exists.

Example:

Simple design demonstrating for loop

```
for (i=0; i<10;i=i+1)
begin
$display ("current loop#%d", i);
end
```

This loop repeats 10 times and each time displays current loop#0 upto 9 times with increment in 'i'.

6.5.5.1 While-Loop Statement

The syntax of while-loop is as follows:

while *(condition)*
begin
procedural_statement;
end

The while-loop executes the procedural statements until the condition becomes false. If the condition results in x or z, then it is treated as 0.

Example:

Simple design demonstrating while-loop

```
While (i<10)
begin
y=y+1;
i=i+1;
end
```

6.5.5.3 Forever-Loop Statement

The syntax for this loop statement is as follows:

forever
begin
procedural_statements;
end

This loop without any condition repeatedly executes, and the disable statement to stop the loop may be written in the procedural assignment. Otherwise, the forever loop will repeat forever.

Example:

Simple design considering the forever loop

```
initial
begin
clk=0;
forever
#20 clk=~clk;
end
```

In this design, the output is the square wave with a period of 40 time units. That means for every 20 time units, the 'clk' toggles its value.

6.5.6 EXAMPLES

A. *Verilog program for positive edge-triggered JK flip-flop*

```
module jk_ff (j,k,clk,q);
input j;
input k;
input clk;
output reg q;
  always @ (posedge clk)
    case ({j,k})
      2'b00:  q <= q;
      2'b01:  q <= 0;
      2'b10:  q <= 1;
      2'b11:  q <= ~q;
    endcase
endmodule
```

B. *Verilog program for 4-bit up counter*

```
module up_counter(clk, reset, output, counter);
input clk, reset;
```

```
output reg [3:0] counter;
reg [3:0] counter_up;
always @(posedge clk or posedge reset)
begin
if(reset)
counter_up <= 4'd0;
else
counter_up <= counter_up + 4'd1;
end
assign counter = counter_up;
endmodule
```

6.6   TASKS AND FUNCTIONS

This section explains both tasks and functions with examples. First, we start with the task, and then we explain the function. Finally, the difference between the task and the function is also explained [4].

6.6.1   TASK

Task is a block of procedural statements (task definition) that can be called from different places of design description. A task can be called at any place of the program and it can call other tasks and functions as well. The definition of task is shown below:

task *task_name*;
procedural_statements;
endtask

We need to write the task definition within the module declaration, and the task can be called by using the below statement:

task_name (argument1, argument2...., argumentN);

List of argument in the task calling should match with the input and output order in the task definition.

Example:

Write a simple program using task (conversion of temperature from Celsius to Fahrenheit)

```
module task_example (temp_a, temp_b, temp_c, temp_d);
input [7:0] temp_a, temp_c;
output [7:0] temp_b, temp_d;
reg [7:0] temp_b, temp_d;
```

```
task convert;//task definition
input [7:0] temp_in;
output [7:0] temp_out;
begin
temp_out = (9/5) * (temp_in + 32)
end
endtask

always @ (temp_a)
begin
convert (temp_a, temp_b);//task calling
end
always @ (temp_c)
begin
convert (temp_c, temp_d);//task calling
end
endmodule
```

6.6.2   FUNCTION

Function is a block of procedural statements like task and can be called from any part of the module. The main difference between task and function is that function can return one value, whereas task cannot return. Function cannot call other task and it must have at least one variable in definition. No output declarations are allowed in function and it may call other functions [5]. The syntax of the function is as follows:

function [range or type] function_name;
input declaration;
procedural_statements;
endfunction

Example:

Function definition for the sum of two variables

```
function [5:0] sum;
input [5:0] x, y;
begin
sum = a+b;
end
endfunction
```

The syntax of function call is shown below:

function_name (argunent1, argument2,.......argumentN);

Example:

Calling a function

```
module function_calling (result, a, b);
output reg [5:0] result;
input [5:0] a, b;
function [5:0] sum; //function definition

begin
sum = a+b;
end
endfunction
always@(a or b)
begin
result= sum(a, b); //function calling
end
endmodule
```

6.7 SUMMARY

This chapter presented the Verilog programming and the methodology to design several digital systems using this language.

- First, the basics and different elements of Verilog were introduced.
- The different modeling styles like gate-level, data-flow, and behavior-level modeling styles were discussed in detail with examples.
- Finally, the used Verilog tasks and functions were also introduced and explained with suitable examples.

6.8 MULTIPLE-CHOICE QUESTIONS

1. HDL is the abbreviation for
   a. Hardware description language
   b. Hardware design language
   c. High description language
   d. Human description language

2. Dataflow modeling mainly uses
   a. Procedural assignments
   b. Blocking assignment
   c. Continuous assignments
   d. Assign statements

3. Logical and is represented as
   a. &&
   b. ||

   c. &
   d. |

4. Compiler directives start with
   a. Backquote
   b. Dot
   c. Semicolon
   d. Equal

5. Which one of the following statements executes only once?
   a. Initial
   b. Always
   c. Repeat
   d. for

6. Which of the following is the higher-level modeling style?
   a. Gate-level
   b. Dataflow level
   c. Behavioral-level
   d. None

7. Which of the following represents concatenation operation?
   a. {}
   b. ≪
   c. ~
   d. []

8. Module in Verilog represents
   a. Design description
   b. Test patterns
   c. Loops
   d. Conditions

9. Module and respective test bench links with
   a. Instantiation
   b. Input and output declaration
   c. Initial statement
   d. Always statement

10. Intermediate connections in dataflow style can be declared as
   a. Wires
   b. Registers
   c. Variable
   d. None of the above.

6.9   SHORT ANSWER QUESTIONS

1. What are Verilog Operators? Explain.
2. What are the different data types in Verilog?
3. What are compiler directives in Verilog?
4. Write the syntax of conditional operator.
5. Write the syntax of 'for loop' in Verilog.
6. What is the task in Verilog?
7. What is the difference between Task and Function in Verilog?
8. What is the syntax of the always statement in Verilog?
9. What is the difference between initial and always statements?
10. Write a Verilog encoder program in dataflow style.

6.10   LONG ANSWER QUESTIONS

1. Explain dataflow style of Verilog modeling with an example.
2. Explain in detail about tristate gates in Verilog.
3. What is module and test bench in Verilog? Explain.
4. What are different condition statements in Verilog? Explain with examples.
5. Explain gate-level style of Verilog modeling with an example.
6. What are the different loops available in Verilog? Explain with examples.
7. Explain behavioral style of Verilog modeling with an example.
8. Explain initial and always statements with examples.
9. Write 4-bit up-down counter program with behavioral-level modeling.
10. Write 4-bit ALU program using behavioral-level modeling.

REFERENCES

1. Bhasker, J. 1998. *Verilog HDL Synthesis: A Practical Primer.* Star Galaxy Publishing, Allentown, PA 18103.
2. Palnitkar, S. 2003. *Verilog HDL: A Guide to Digital Design and Synthesis* (Vol. 1). Prentice Hall Professional, Upper Saddle River, New Jersey.
3. Cavanagh, J. 2017. *Verilog HDL: Digital Design and Modeling.* CRC Press, Boca Raton, Florida
4. Minns, P.D. and Elliott, I.D. 2008. *FSM-based digital design using Verilog HDL.* J. Wiley & Sons, Hoboken, New Jersey, United States.
5. Sutherland, S, 2013. *The Verilog PLI Handbook: a User's Guide and Comprehensive Reference on the Verilog Programming Language Interface.* Springer Science & Business Media, Berlin, Germany.

7 VLSI Interconnect and Implementation

7.1 AN OVERVIEW OF THE VLSI INTERCONNECT PROBLEM

Very-large-scale integration (VLSI) industry has tremendous growth over several decades due to scaling the devices and interconnect by fulfilling the demand and requirement. According to Moore's law, the feature size of integrated circuits (ICs) reduces and the total number of transistors rapidly increases, which gets double on a single silicon chip after every 2 years [1]. The designer is continuously following Moore's concept; however, the complexity of the electronics system increases with the innovations and demand for high-speed operations. In the early stage of the ICs industry, the speed of ICs depended on the gate delay instead of the interconnect delay [2]. Consequently, the interconnect is considered a second-class citizen that appears only at the time of any high-performance task or any special computational task. As the technology has moved forward from deep submicron to nanotechnology, several reductions have been demonstrated such as the size of the metal oxide semiconductor (MOS) transistor, silicon area, and power consumption. In order to scale down the technology and operate at high frequency, the impact of interconnect needs to be considered that primarily dominates the performance of overall on-chip ICs. Therefore, several materials have been used as an interconnect application such as aluminum (Al), gold (Au), silver (Ag), and copper (Cu) that have their own limitations over technology scaling [3].

7.1.1 INTERCONNECT SCALING PROBLEM

In a typical CMOS VLSI technology, the area occupied by the active device is approximately 10%, while it is 6–10 times more for interconnects. Consequently, the importance of interconnects achieved major attention as the device density increased and the feature size scaled down. In the early stages of the ICs industry, Al was preferred as interconnect metal because of its low resistivity, good adherence to silicon (Si), boundability, patternability, and ease of deposition [4]. In addition, it causes negligible contamination with undesirable impurities in the ICs, and it is easily available as low-cost material in the earth. However, despite its several advantages, it tends to have several reliability issues due to electromigration and contact failure effect that mainly has a major impact on the interconnect performance. Electromigration is the result of the movement of atoms from one place to another that is caused by the gradual movement of ions in the conductor [5]. Later, Ag and Au were also investigated for interconnect application, but due to higher electromigration and quick diffusion into silicon during fabrication the use of Al, Au, and Ag materials was restricted. In the late 1990s, Cu changed the silicon industry as it provided better reliability, lower resistivity, higher melting points, higher current density, and ten times more resistance to electromigration as compared to the above-mentioned materials. The higher melting point of Cu provides improved thermal stability due to several advantages of Cu that turn into preferred interconnect material for deep submicron technology. As the VLSI technology is further scaled down, the performance of interconnect is facing serious challenges, especially at global signals. Demand for higher operating speed and frequency led to a gradual increase in the resistivity of interconnect due to electron surface and grain boundary scattering [2]. As the dimensions are reduced rapidly, it becomes comparable to mean free path (MFP) of an electron (~40 nm at room temperature in a Cu), that results in scattering from the interface that cannot be ignored due to surface scattering, and hence the resistivity of copper increases rapidly. Another effect of dimension scaling for Cu interconnect is the increase in current density with technology scaled down. The per unit length (*p.u.l.*) of interconnect parasitics increases, which causes the degradation of interconnect performance and leads to higher propagation delay and more power dissipation. The rise in power dissipation also causes an increase in heating that results in electromigration [6]. As these limitations of Cu interconnect are technology dependent and are going to be more and more severe for the future generation of VLSI chips. Therefore, it is necessary to look forward to an alternative emerging material.

7.1.2 Implementation of Interconnect Problem

To understand the implementation of interconnect, we start with the basics of interconnect and their scaling effect. Based on the length of the wire, the interconnect lines are categorized into three types: local, intermediate, and global interconnects, as shown in Figure 7.1 [6]. Local interconnect is used to connect short-distance components within the functional blocks and occupy first or sometimes second metal layer. Table 7.1 demonstrates different scaling factors for local and global interconnects with associated interconnect parasitic.

It usually connects gates, source, and drain in the metal oxide semiconductor technology. Intermediate interconnect is used to provide clock and distributed

---

TABLE 7.1
Scaling Factor of Local and Global Interconnects

Parameters	Symbol	Scaling Factor
Cross-sectional dimensions	(W, H, S_{i-i}, t_{ox})	$1/S$
Resistance in *p.u.l.*	$R_{p.u.l}$	S^2
Capacitance in *p.u.l.*	$C_{p.u.l}$	1
RC constant in *p.u.l.*	$R_{p.u.l} \cdot C_{p.u.l}$	S^2
Local interconnect length	l_{loc}	$1/S$
Local interconnect *RC* delay	$R_{p.u.l} \cdot C_{p.u.l} \cdot l_{loc}^2$	1
Die size	D_s	S_s
Global interconnect length	l_{glo}	S_s
Global interconnect *RC* delay	$R_{p.u.l} \cdot C_{p.u.l} \cdot l_{glo}^2$	$S^2 S_s^2$

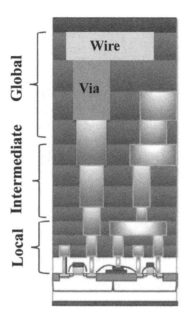

FIGURE 7.1 Type of interconnects.

signal and to connect devices within a functional block. Intermediate interconnect is also referred to as a semiglobal interconnect and designed slightly wider
and taller than local interconnect to provide lower resistance because the resistance reduces with more area. Global interconnect is used to provide clock and
distributed signals between the functional block and to provide power/ground
to all functional blocks. Global interconnect primarily occupies the top one
or two layers, and it should have lower resistance as well as less propagation
delay because the overall performance of the chip mainly depends on global
interconnect.

The reason for the ever-increasing importance of interconnect can be understood
using the electrical models that are fully justified in Subsection 7.3; before that, let
us have some introductory explanation. The implementation of interconnect into
its equivalent electrical model has a major impact due to changes in its parasitic
quantitative values based on the technology node. One can also understand that the
delay is mostly dependent on the resistance (R) and capacitance (C); apart from this,
inductance (L) is also important for the high-frequency signal. At the same time,
the increase in the device density leads to reduced cross-sectional dimensions that
increase *p.u.l.* parasitics of interconnect and degrades the overall performance of the
VLSI chip.

  i. **Increase in complexity:** In the 1980s, the impact of chip technology was
     studied to understand the impact of delay on-chip performance. It was shown
     that the increase in complexity implies an increase in the length of global
     interconnects that increases the delay. This increase in the interconnect

length can be expressed as a function of the overall area of the chip for the global interconnect as:

$$l \propto \sqrt{A} \qquad (7.1)$$

where l and A are referred to as the length and chip area of global interconnect.

ii. **Signal frequency:** Devices (transistor) driving capability increases as a result of technology scaling. This means higher frequency components in signals are transmitted through interconnects. A simple analysis can give a dependence between the rise/fall time of a digital signal transition and its maximum significant frequency, F_{max}:

$$F_{max} \approx \frac{0.5}{t_r} \qquad (7.2)$$

where t_r is the rise time of the signal. Higher frequency content means two factors: First, the circuit parameters L and C that have a greater impact on the line impedance and its effect is proportional to the frequency. Second, a reflection of the signal while operating at a higher frequency signal.

7.2   INTERCONNECT AWARE DESIGN METHODOLOGY AND ELECTRICAL MODELING

In this section, the reader can understand the impact of scaling from the early stage of MOS technology to advanced technology for the transistor and interconnect level.

7.2.1   IMPACT OF SCALING

In the early stage of VLSI technology, the impact of scaling was only limited to device parameters due to lower speed and frequency. The area of silicon chip reduces generation by generation due to improvement in the lithography technique and process technology. The journey from micron technology to deep submicron technology is shown in Figure 7.2, which demonstrates almost 30% improvement in the technology after every 2–3 years while following Moore's law [1]. With the advancement of technology, the size of transistors in a single chip becomes smaller and exhibits several performance improvements such as faster switching, lesser power consumption, and cheaper to manufacture in bulk quantity. Further scaling in the feature size generates several fabrication challenges, noise, and reliability issues. Therefore, the designer needs to take care of all the chip performance aspects before executing for future production.

7.2.2   TRANSISTOR SCALING

The model formulated by **Dennard** is used for scaling the first-order constant field MOS technology in the early stage of the VLSI industry. The scaling of the transistor

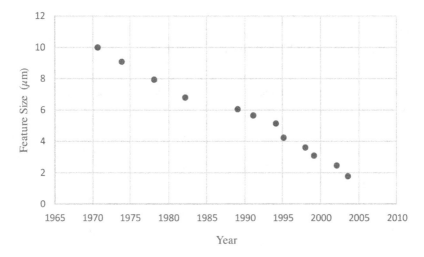

FIGURE 7.2   Reduction in feature size with the change in generation [6].

can be done in several ways, wherein the critical parameters of the MOS device are scaled by factor S to preserve the basic operational by maintaining the characteristics of the MOS device [7–11]. The scaling parameters are as follows:

1. Entire dimensions (in all coordinates x, y, and z)
2. The voltage level of device
3. Doping concentration densities

Apart from this approach, the *lateral scaling* is also used by scaling the channel (gate) length only, while other dimensions, voltages, and doping concentration level remain unaltered; that is why it is mostly called gate shrink.

In the early stage of VLSI technology, the *constant voltage scaling* was maintained as 5 V by reducing the feature size from 6 μm to 1 μm that led to a reduced cost and quadratic improvement in the delay. The constant voltage scaling no longer sustains below 1 μm technology due to increased electric fields in MOS devices. Hence, the devices are at risk and breakdown due to the high field.

In addition to that, the *constant field scaling* is used by scaling all the dimensions as illustrated in Table 7.2, including width W, the channel length L, thickness of oxide t_{ox}, and supply voltage V_{dd} shrunk by a factor of $1/S$.

In addition to that, scaling of metal oxide semiconductor field-effect transistor (MOSFET) is categorized into three types: Constant field scaling, Constant voltage scaling, and Lateral scaling.

1. **Constant field scaling**: In this category of scaling, the designer preserves the amplitude of the internal electric field in the MOSFET, while the other

TABLE 7.2

Impact of Scaling on MOS Device Characteristics

Parameters	Symbol	Constant field	Constant voltage	Lateral
		Scaling parameters		
Channel width	W	$1/S$	$1/S$	1
Channel length	L	$1/S$	$1/S$	$1/S$
Substrate doping	N_A	S	S	1
Oxide thickness of gate	t_{ox}	$1/S$	$1/S$	1
Supply voltage	V_{dd}	$1/S$	1	1
Threshold voltage	V_{tn}, V_{tp}	$1/S$	1	1
Saturation current	I_{ds}	$1/S$	S	S
Effective resistance	$R_{eff} = V_{dd}/I_{ds}$	1	$1/S$	$1/S$
Gate capacitance	$C = (W \cdot L/t_{ox}I)$	$1/S$	$1/S$	$1/S$
Gate delay	$\tau = RC$	$1/S$	$1/S^2$	$1/S^2$
Clock frequency	$f = 1/\tau$	S	S^2	S^2
Dynamic power dissipation	$P = C \cdot V^2 \cdot f$	$1/S^2$	S	S
Chip area	A	$1/S^2$	$1/S^2$	1
Power density	$P_{density} = P/A$	1	1	S
Current density	$I_{density} = I_{ds}/A$	S	S^3	S

dimensions are scaled down by a factor of S. The impact of this kind of scaling on different factors is described below:

a. **Scaling of gate oxide capacitance per unit area:** The current-voltage characteristics of the MOS transistor is now considered based on full scaling while assuming that the surface mobility does not get affected by scaled doping density. In this case, it can be modified as

$$C'_{ox} = \frac{\varepsilon_{ox}}{t'_{ox}} = S\frac{\varepsilon_{ox}}{t_{ox}} \qquad (7.3)$$

$$C'_{ox} = SC_{ox} \qquad (7.4)$$

b. **Scaling of linear drain current:** The W/L ratio will remain unchanged for scaling the linear drain current of the MOSFET, while the transconductance parameter k_n gets scaled by a factor of S and can be found as

$$I'_D(lin) = \frac{k'_n}{2}\left[2(V'_{GS} - V'_T)V'_{DS} - V'^2_{DS}\right] \qquad (7.5)$$

$$= \frac{S.k'_n}{2}\frac{1}{S^2}\left[2(V_{GS} - V_T)V_{DS} - V^2_{DS}\right] \qquad (7.6)$$

$$I'_D(lin) = \frac{I_D(lin)}{S} \qquad (7.7)$$

Similarly, the saturation-mode current can also be found as

$$I_D'(sat) = \frac{k_n'}{2}\left(V_{GS}' - V_T'\right)^2 = \frac{S.k_n'}{2}\frac{1}{S^2}\left(V_{GS} - V_T\right)^2 \tag{7.8}$$

$$I_D'(sat) = \frac{I_D(sat)}{S} \tag{7.9}$$

c. ***Scaling of power dissipation and power density:*** The instantaneous power dissipation by the device can be observed before scaling, since the drain current flows between the source and the drain node and can be found as

$$P = I_D V_{DS} \tag{7.10}$$

Hence, the power dissipation of the transistor can be reduced by a factor of S^2 because both the drain current and the drain to source voltage are scaled by a factor of S in full scaling.

Therefore,

$$P' = I_D' V_{DS}' = \frac{1}{S^2} I_D V_{DS} \tag{7.11}$$

$$P' = \frac{P}{S^2} \tag{7.12}$$

It can be observed that the power dissipation drastically reduced by a factor of S^2 in full scaling. Similarly, power density is the ratio of power per unit area and can be written before scaling as

$$Power\ density\ (P_D) = \frac{P}{Area} \tag{7.13}$$

After scaling it can be found as

$$P_D' = \frac{P.S^2}{S^2 Area} = \frac{P}{Area} \tag{7.14}$$

$$P_D' = P \tag{7.15}$$

Therefore, the power density remains unchanged in per unit area for the scaled devices.

d. ***Scaling of gate delay or propagation delay:*** As we know that the current flowing through capacitance is

$$I = C_g \frac{dV}{dt} \tag{7.16}$$

Gate delay before scaling as

$$t = C_g \frac{V}{I} \tag{7.17}$$

The gate capacitance (C_g) before scaling is also scaled down by a factor of S

$$C_g = WLC_{ox} \tag{7.18}$$

The gate capacitance after scaling is

$$C'_g = W'L'C'_{ox} = \frac{W}{S}\frac{L}{S}SC_{ox} \tag{7.19}$$

$$C'_g = \frac{C_g}{S} \tag{7.20}$$

Gate delay after scaling as

$$t' = \frac{C'_g V'}{I'} = \frac{C_g}{S}\frac{V}{S}\frac{S}{I} \tag{7.21}$$

$$t' = \frac{t}{S} \tag{7.22}$$

Therefore, the gate delay is scaled down by a factor of S.

2. **Constant voltage scaling:** In constant voltage scaling, the supply voltage is kept constant while the other physical and process parameters are scaled by a factor of S as illustrated in Table 7.2.

   a. *Scaling of gate oxide capacitance per unit area:* The gate oxide capacitance can be observed in per unit area before scaling as

$$C_{ox} = \frac{\varepsilon_{ox}}{t_{ox}} \tag{7.23}$$

Similarly, after scaling

$$C'_{ox} = \frac{\varepsilon_{ox}}{t'_{ox}} = S \cdot \frac{\varepsilon_{ox}}{t_{ox}} = S \cdot C_{ox} \tag{7.24}$$

$$C'_{ox} = S \cdot C_{ox} \tag{7.25}$$

Here, the gate capacitance per unit area is increased by a factor of S.

   b. *Scaling of transconductance:* The transconductance before scaling can be represented as

$$k_n = \mu_n C_{ox}\left(\frac{W}{L}\right) \tag{7.26}$$

Similarly, it can be written after scaling as

$$k'_n = \mu_n C'_{ox}\left(\frac{W'}{L'}\right) = S \cdot \mu_n C_{ox}\left(\frac{W}{L}\right) \tag{7.27}$$

$$k'_n = S \cdot k_n \tag{7.28}$$

Here, the transconductance is increased by a factor of S.

c. ***Scaling of linear mode drain current:*** The linear mode drain current before scaling can be found as

$$I_D(lin) = \frac{k_n}{2}\left[2(V_{GS} - V_T)V_{DS} - V_{DS}^2\right] \tag{7.29}$$

Similarly, after scaling, the terminal voltage get unchanged and can be written as

$$I_D'(lin) = \frac{k_n'}{2}\left[2(V_{GS} - V_T)V_{DS} - V_{DS}^2\right] = S \cdot \frac{k_n}{2}\left[2(V_{GS} - V_T)V_{DS} - V_{DS}^2\right] \tag{7.30}$$

$$I_D'(lin) = S \cdot I_D(lin) \tag{7.31}$$

Here, the linear mode drain current is increased by a factor of S.

d. ***Scaling of Saturation mode drain current:*** The saturation drain current before applying the scaling can be found as

$$I_D(sat) = \frac{k_n}{2}\left[(V_{GS} - V_T)^2\right] \tag{7.32}$$

Similarly, after scaling, it can be found as

$$I_D'(sat) = \frac{k_n'}{2}\left[(V_{GS} - V_T)^2\right] = S \cdot \frac{k_n}{2}\left[(V_{GS} - V_T)^2\right] \tag{7.33}$$

$$I_D'(sat) = S \cdot I_D(sat) \tag{7.34}$$

Here, the saturation mode drain current is increased by a factor of S.

e. ***Scaling of power dissipation and power density:*** The power dissipation before scaling can be found as

$$P = I_D \cdot V_{DS} \tag{7.35}$$

After scaling, it can be found as

$$P' = I_D' \cdot V_{DS} = S \cdot I_D \cdot V_{DS} \tag{7.36}$$

$$P' = S \cdot P \tag{7.37}$$

Here, the power dissipation is increased by a factor of S.
Similarly, the power density before scaling can be found as

$$P_D = \frac{P}{Area} \tag{7.38}$$

Power density after scaling,

$$P_D' = \frac{P'}{Area} = \frac{S.P}{\frac{Area}{S^2}} = S^3 \frac{P}{Area} \qquad (7.39)$$

$$P_D' = S^3 P_D \qquad (7.40)$$

Therefore, it can be observed that the power density drastically increases by a factor of S^3.

3. **Lateral scaling:** Similarly, the lateral scaling factor for different physical parameters can be identified wherein only the gate length is scaled by a factor of S, as shown in Table 7.2. The reader can now have a deep knowledge about the different scaling approaches that are used in the VLSI domain.

Of these three scaling techniques, the constant voltage scaling is mostly preferred over the constant field and lateral scaling in many cases, while the supply voltages and all the node voltages are scaled proportionally with the dimension of the device in full scaling technique. The constant voltage scaling is mostly used to reduce the multiple power supply voltage at a certain voltage level that is required for several internal and external peripherals that make them complicated and more complex architecture. To overcome these issues, the contact voltage scaling is usually preferred instead of full scaling or constant field scaling. However, it can also be observed from Table 7.2 that using constant voltage scaling in a MOS transistor, the drain current and power dissipation increases by a factor of S, while power dissipation density increases by a factor of S^3. This large increase in power dissipation density may cause serious reliability issues such as electromigration, hot carrier degradation, oxide breakdown, and electrical overstress for transistor scaling.

7.2.3 INTERCONNECT SCALING

In general, interconnect scaling can be obtained in two ways: either scale all the dimensions or keep the wire height constant. The detailed description of interconnect scaling by a scaling factor of S is illustrated in Table 7.3. In high-performance VLSI technology, the impact of interconnect is considered that has been neglected in the early stage of VLSI technology due to the lower operating speed and performance and hence only the gate delay is estimated. In the era of advanced submicron technology, the on-chip interconnect primarily dominates the overall system performance. Therefore, interconnect plays an important role in designing high-speed ICs.

7.3 ELECTRICAL CIRCUIT MODEL OF INTERCONNECT

For several decades, the evaluation of circuit modeling has developed to understand the physical behavior of interconnect in terms of speed and technology scaling. Therefore, circuit modeling is categorized as an ideal interconnect, resistive interconnect, capacitive interconnect, and resistive tree interconnect, as shown in Figure 7.3 [12].

TABLE 7.3

Impact of Scaling on Interconnect Characteristics

		Scaling Parameters	
Parameters	Symbol	Reduced Thickness	Constant Thickness
Width	W	$1/S$	$1/S$
Thickness	T	$1/S$	1
Spacing	S_p	$1/S$	$1/S$
Dielectric height	h_g	$1/S$	$1/S$
Per Unit Length (p.u.l.) Characteristics			
Wire resistance	$R_w = 1/W{\cdot}t$	S^2	S
Fringing capacitance	$C_{fc} = t/S_p$	1	S
Parallel plate capacitance	$C_{pp} = W/h_g$	1	1
Total wire capacitance	$C_{tw} = C_{fc}+C_{pp}$	1	Between 1, S
Unrepeated RC constant	$R_w{\cdot}C_w$	S^2	Between S, S^2
Crosstalk noise	t/S_p	1	S

7.3.1   IDEAL INTERCONNECT

In the early stage of the VLSI technology era, the high-speed operation and large power dissipation were not major concerns. However, the performance was dominated by gate and active device delay instead of interconnect delay. Hence, the impact of resistance and capacitance was neglected using ideal electrical interconnect lines between the elements.

7.3.2   RESISTIVE INTERCONNECT

In the late 1980s and 1990s, electrical modeling became more complex and the impact of resistance became non-negligible as compared with effective resistance driving transistors. The general wire geometry to estimate the resistance is depicted in Figure 7.4, which consists of an effective resistivity ρ_x based on the material, as shown in Table 7.4. The interconnect wire shows resistive behaviors having width W_x, thickness T_x, and length L_x, wherein the interconnect resistance is proportional to its wire length and inversely proportional to the cross-sectional area A_x. The resistance (R_x) of uniform rectangular wire can be calculated as

$$R_x = \frac{\rho_x L_x}{A_x} = \frac{\rho_x L_x}{T_x W_x} \tag{7.41}$$

$$R_x = R_{sqr}\frac{L_x}{W_x} \tag{7.42}$$

where $R_{sqr} = \rho_x/T_x$ is a process-dependent parameter referred to as the sheet resistance or resistance exhibit by per square in Ω/square.

TABLE 7.4
Commonly Used Resistivity of Interconnect Material at 22°C

Interconnect Material	Symbol	The Quantitative Value of Resistivity ρ_x ($\mu\Omega$ cm)
Silver	Ag	1.6
Copper	Cu	1.7
Gold	Au	2.2
Aluminum	Al	2.8
Molybdenum	Mo	5.3
Tungsten	W	5.5
Titanium	Ti	43.0

(a) Ideal Interconnect

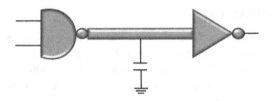

(b) Capacitive Interconnect

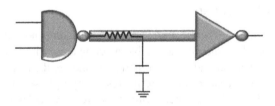

(c) Resistive Interconnect

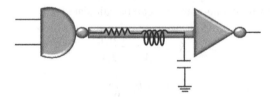

(d) Inductive Interconnect

FIGURE 7.3   Different scenarios of interconnect modeling.

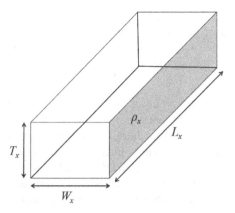

FIGURE 7.4 Basic rectangular wire segment.

7.3.3 CAPACITIVE INTERCONNECT

The study of interconnect capacitance is the subject of advanced research and hence accurate modeling of interconnect structure is still complicated due to three-dimensional integrated circuits. The interconnect capacitance primarily depends on its distance to the substrate, between the wires, distance from the surrounding metal wire, and the shape of interconnect. Rather struggling with the complicated interconnect geometrical structure, the designer extracts the interconnect capacitance using an industry-standard advanced extraction tool to achieve precise quantitative values from the layout.

In the late 1970s and 1980s, interconnect capacitance was extracted as area capacitance to substrate for single-layer metallization. Later on, the impact of fringing capacitance has been included in the modeling of interconnect, as shown in Figure 7.5.

For a simple rectangular wire, the overall capacitance of interconnect can be modeled as parallel plate capacitance (C_{p2p}), if the wire width is larger than the thickness of the dielectric material while assuming that the electric field lines are orthogonal to

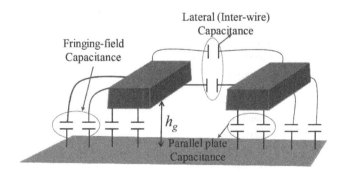

FIGURE 7.5 The overall impact of capacitance.

capacitor plate [13, 14]. The parallel plate capacitance is also known as area capacitance and the capacitance can be calculated with the general expression as

$$C_{p2p} = \frac{\varepsilon_{r,die}}{T_x} W_x L_x \tag{7.43}$$

The main concern of expressing equation (7.43) is to make the reader understand the effect of dielectric material that is proportional to the interconnect capacitance. Practically, the above model is too simplistic and cannot be considered further to estimate the accurate analysis of overall interconnect capacitance. From the fabrication point of view, scaling down the technology node having a lower dimension of width creates denser proximity of wire and less area overhead. Due to those circumstances, the simplest form of equation (7.43) no longer provides an accurate measurement of capacitance. Therefore, fringing capacitance needs to be considered that are generated due to the sidewall of the wires by considering the exact model of complex geometry. The overall capacitance for the exact model is approximated as the sum of the parallel plate and fringing capacitance that can be expressed as

$$C_t = C_{p2p} + C_{fringe} \tag{7.44}$$

$$C_t = \frac{\varepsilon_{r,die}}{T_x} W_x L_x + 2\varepsilon_{r,die} \ln\left(1 + \frac{T_x}{h_g}\right) \tag{7.45}$$

To understand the concept of extracting the accurate measurement of interconnect capacitance, the designer needs to consider all the possible combinations of capacitance that occur between the wires placed above the semiconductor substrate. The relative permittivity of some of the dielectric substrate material used in IC is illustrated in Table 7.5. The detailed understanding of estimating the interconnect capacitance can be observed in Subsection 7.4.2.

TABLE 7.5
Relative Permittivity of Commonly Used Dielectric Materials

Dielectric Material	$\varepsilon_{r,die}$
Free space	1
Aerogel	1.5
Polyimides (organic)	3–4
Silicon dioxide	3.9
Glass-epoxy (PC board)	5
Silicon Nitride	6.5–7.5
Alumina	9.5
Silicon	11.7
Zirconium oxide	23.0
Hafnium oxide	20.0
Tantalum oxide	20–30

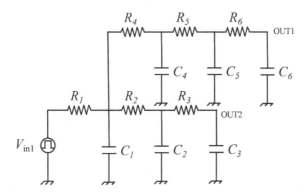

FIGURE 7.6   A general RC network consisting of a routing tree.

7.3.4   RESISTIVE INTERCONNECT TREE

In general, the wire is considered the resistive net that corresponds to the routing tree. These routing trees can be modeled as the RC tree network, as shown in Figure 7.6. In the case of the RC tree network, the following points can be observed:

i. All the capacitors connected in the RC network are grounded.
ii. Only one acting input can be observed.
iii. The routing path has no resistive loop.

To measure the delay of the RC network, it requires a very large and complex differential equation. To obtain the delay, one can use a simple model based on Elmore that is equivalent to the first moment of the waveform [14–16]. The Elmore delay model provides a simple and fast calculation of delay.

According to the Elmore delay, the interconnect tree delay can be found from source to any ith node in the tree and can be expressed as

$$\tau_{Elmore} = \sum_{S=1}^{i} C_s \sum_{\substack{for\ all\ path \\ k \in P_{sk}}} R_k \qquad (7.46)$$

For example, the Elmore delay at node OUT2 can be found based on expression (7.46) as

$$\tau_{Elmore} = R_1 C_1 + R_1 C_4 + R_1 C_5 + R_1 C_6 + (R_1 + R_2)C_2 + (R_1 + R_2 + R_3)C_3$$

Similarly, the Elmore delay can also be calculated at node OUT1 as

$$\tau_{Elmore} = R_1 C_1 + R_1 C_2 + R_1 C_3 + (R_1 + R_4)C_4 + (R_1 + R_4 + R_5)C_5 + (R_1 + R_4 + R_5 + R_6)C_6$$

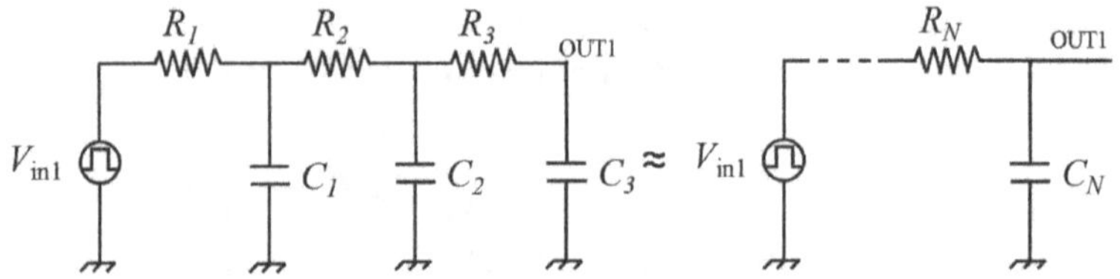

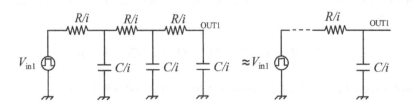

FIGURE 7.7 Distributed *RC* network tree consisting of one branch.

For a special case of *RC* network, the cascade arrangement, shown in Figure 7.7 is simplified as an equivalent single branch and its corresponding Elmore delay from the input node to Node *N* can be calculated as

$$\tau_{Elmore,N} = \sum_{s=1}^{i} C_s \sum_{k=1}^{s} R_k \qquad (7.47)$$

Similarly, if the *RC* network tree consists of a series combination of resistance (*R/i*) and capacitance (*C/i*), as shown in Figure 7.8, then the Elmore delay can be calculated as

$$\tau_{Elmore,i} = \sum_{s=1}^{i} \left(\frac{C}{i}\right) \sum_{k=1}^{s} \left(\frac{R}{i}\right)$$

$$= \left(\frac{C}{i}\right)\left(\frac{R}{i}\right)\left(\frac{i(i+1)}{2}\right) = RC\left(\frac{(i+1)}{2i}\right) \qquad (7.48)$$

For very large distributed *RC* network tree, the above delay expression can be reduced as

$$\tau_{Elmore,i} = \left(\frac{RC}{2}\right) \; for \; i \to \infty \qquad (7.49)$$

It can be observed that the delay of the distributed *RC* network line is considerably lower than that of the lumped *RC* network.

FIGURE 7.8 Distributed *RC* network consisting of *N* equal segment.

7.4 ESTIMATION OF INTERCONNECT PARASITICS

This section provides a detailed calculation of interconnect parasitic that is used to understand the performance in terms of propagation delay, crosstalk effect, power dissipation, and so on.

Here, we begin with the calculation of resistance, inductance, and capacitance associated with interconnect by assuming a rectangular geometrical structure.

7.4.1 INTERCONNECT RESISTANCE ESTIMATION

Interconnect used in VLSI has different resistance based on the resistivity of metal used for interconnect applications at different technology nodes. Scaling of technology has a major impact on resistance value due to a reduced width and thickness. The basic model of interconnect resistance is shown in Figure 7.4. The resistance R_x can be described as follows:

$$R_x = \frac{\rho_x L_x}{A_x} = \frac{\rho_x L_x}{T_x W_x}$$

$$= \frac{L_x}{\sigma_x T_x W_x} = \frac{L_x}{(e\mu n)T_x W_x} \tag{7.50}$$

$$R_x = R_{sqr} \frac{L_x}{W_x} \tag{7.51}$$

where ρ_x is the resistivity of conducting wire that depends on the charge of electron e, mobility μ, and carrier concentration n, referred to $\sigma_x = e\mu n$. The number of segmented squares of the conducting metal wire is defined as the ratio of length and width [3]. Therefore, depending on the above facts, the square resistance of wire having length L_x is termed as the sheet resistance that is defined as the ratio of resistivity to the thickness of the material as

$$R_{sqr} = \frac{\rho_x}{T_x} \; \Omega/\square \tag{7.52}$$

Using equation (7.51), we can also say that the interconnects having different width and length but the same ratio can provide the same resistance as offered by equation (7.51) that can be observed from below instances:

For structure (a), as shown in Figure 7.9, the overall resistance can be calculated as

$$R_x = R_{sqr} \frac{L_x}{W_x} \Omega \tag{7.52a}$$

For structure (b), as shown in Figure 7.9, the overall resistance can be calculated as

$$R_x = R_{sqr} \frac{2L_x}{2W_x} = R_{sqr} \frac{L_x}{W_x} \Omega \tag{7.52b}$$

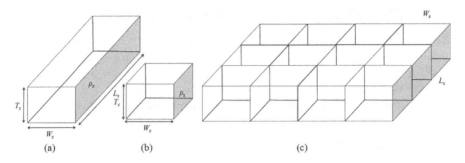

FIGURE 7.9 Resistance estimation for different structures.

Therefore, it can be concluded that both the structures (a) and (b) have the same resistance. Figure 7.10 illustrates the pitch that can be defined as the distance between the center of the width for the coupled line. It can be defined as $p = S_x + W_x$ and related parameter based on 6-metal 180 nm technology is listed in Table 7.6. Based on the International Technology Roadmap for Semiconductors (ITRS) [17], it should be noted that the width of the wire is equal to the spacing between the coupled interconnects [18].

Example 7.1:

Calculate the sheet resistances of Copper (Cu) having a metal 1 wire thickness of 0.48 μm and metal 5 wire having a thickness of 1.6 μm.

SOLUTION:

As we can observe from Table 7.4, the resistivity of Cu ≈ 1.7 μΩ cm, therefore using equation (7.52)
The sheet resistance for metal 1,

$$R_{sqr} = \frac{\rho_x}{T_x} = \frac{1.7 \text{ μΩ cm}}{0.48 \text{ μm}} = 35.42 \text{ mΩ}/\square$$

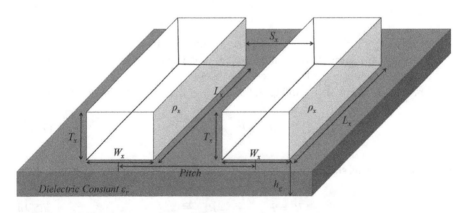

FIGURE 7.10 Interconnect structure placed above the dielectric surface.

TABLE 7.6

Parameters Used for 180 nm Process Technology for Metal-6

Layer	W_x (nm)	T_x (nm)	h_g (nm)	S_x (nm)	Aspect Ratio
1	250	480	800	250	1.9
2	320	700	700	320	2.2
3	320	700	700	320	2.2
4	540	1080	700	540	2.0
5	800	1600	1000	800	2.0
6	860	1720	1000	860	2.0

Similarly, for metal 5,

$$R_{sqr} = \frac{\rho_x}{T_x} = \frac{1.7 \ \mu\Omega \ cm}{1.6 \ \mu m} = 10.625 \ m\Omega / \square$$

From this example, one can easily understand that for longer interconnects, the impact of resistance cannot be simply ignored just because of the small value of sheet resistance in $m\Omega/\square$ as ignoring these will give an inaccurate analysis of the system.

Example 7.2:

Assuming that metal 1 wire having width and length of 1.2 μm and 0.007 m, respectively. Find out the resistance of metal 1 made of Cu.

SOLUTION:

The resistance of metal 1 made of Cu can be calculated as

$$R_{Cu} = R_{sqr} \frac{L_x}{W_x} \Omega$$

$$R_{Cu} = R_{sqr} \frac{L_x}{W_x} = \left(35.42 \times 10^{-3}\right) \frac{0.007}{1.2 \times 10^{-6}} = 206.62 \ \Omega$$

7.4.2 INTERCONNECT INDUCTANCE ESTIMATION

At high operating frequency and lower technology node, the accurate estimation of inductance is essential in order to check the overall system performance. The inductance appears when the current flows through the conducting wire that produces electromagnetic flux around and stored the energy. According to Faraday's law, the voltage induced across the inductive element can be observed, when there is a change in the flow of current on conducting wire and can be expressed as

$$\Delta V = L \frac{di}{dt} \tag{7.53}$$

The inductance and resistance values also get affected by skin and proximity effect due to change in the current flowing through conducting wire caused by an increase in the frequency. In addition to that, the inductance of the wire that depends on the width, thickness, and length can be calculated as

$$L = \frac{\mu_0}{2\pi}\left[L_x \ln\left(\frac{2l}{w+t}\right) + \frac{L_x}{2} + 0.2235(w+t)\right]$$ (7.54)

Example 7.3:

At 180 nm technology, calculate the overall inductance of a 1000 μm metal 1 wire having width and thickness of 0.6 μm and 1.32 μm, respectively. Consider the inductance of a metal 1 wire above the field oxide of thickness 0.53 μm.

SOLUTION:

Using equation (7.54), we can find the inductance of wire as

$$L = \frac{4\pi \times 10^{-13}}{2\pi}\left[1000 \ln\left(\frac{2\times 1000}{0.6+1.32}\right) + \frac{1000}{2} + 0.2235(0.6+1.32)\right]$$

$$L = 2\times 10^{-13}[6948.577 + 500.89502] = 1.4899 \text{ pH}$$

Inductance effect: In a transmission line, the impact of inductance and capacitance has to be considered and that can be identified based on the criterion: If the different two node values are calculated along the wire, then the inductance should be considered and the wire is termed as transmission line, otherwise it can be ignored.

On the other hand, one can also understand that when we need to consider the inductive effect on the transmission line by estimating the rise time at the output of the driving buffer. In this regard, if the round trip time of the signal is lesser than the rise time, then the inductive effect can be neglected [19, 20]. Consequently, the wire length has a lower bound above which the wire inductance should be considered. Based on the above statement, it can be related as

$$\frac{t_r}{2\sqrt{LC_x}} \le L_x \le \frac{2}{R_x}\sqrt{\frac{L}{C_x}}$$ (7.55)

where $v = 1/\sqrt{LC_x}$ is denoted as signal velocity and $Z_0 = \sqrt{L/C_x}$ is referred to as the characteristics impedance of the wire. In some of the case, this range can be nonexistence based on the following condition

$$t_r > 4\frac{L}{R_x}$$ (7.56)

This shows that under these conditions, the impact of inductance is not having an effect for any length.

Example 7.4:

Assume that the 0.8 μm width of a wire exhibits a sheet resistance of 68 mΩ/□, per unit length capacitance of 0.12 fF/μm, and per unit length inductance of 0.4 pH/μm, respectively. Calculate the range of wire length in which the inductance needs to consider if

1. The signal rise time is 2 ns.
2. The signal rise time of 18 ps.

SOLUTION:

1. For a give rise time of 2 ns, the range of wire length as per equation (7.56) can be obtained as

$$t_r > 4 \frac{L}{R_x}$$

$$2 \text{ ns} > 4 \frac{0.4 \text{ pH/μm}}{(68 \text{ mΩ/□})(1□/0.8 \text{ μm})} = 0.02 \text{ ns}$$

Hence, we can observe that for any length of wire, the consideration of wire is not important.

2. For a given rise time of 18 ps, using equation (7.56)

$$0.018 \text{ ns} > 4 \frac{0.4 \text{ pH/μm}}{(68 \text{ mΩ/□})(1□/0.8 \text{ μm})} = 0.02 \text{ ns}$$

Hence, it does not obey the inequality equation, In this regard, we need to consider the impact of inductance for the given length range of the wire. Using equation (7.55), we can conclude that the interconnect inductance is important when length falls between the following two ranges:

$$\frac{t_r}{2\sqrt{LC_x}} \le L_x \le \frac{2}{R_x}\sqrt{\frac{L}{C_x}}$$

$$\frac{0.018 \text{ ns}}{2\sqrt{0.4 \frac{\text{pH}}{\text{μm}} \times 0.12 \text{ fF/μm}}} \le L_x \le \frac{2}{80 \text{ mΩ/μm}}\sqrt{\frac{0.4 \frac{\text{pH}}{\text{μm}}}{0.12 \text{ fF/μm}}}$$

$$1.29 \text{ mm} \le L_x \le 1.45 \text{ mm}$$

Thus, the final wire length range varies between $1.29 \text{ mm} < L_x < 1.45 \text{ mm}$.

7.4.3 INTERCONNECT CAPACITANCE ESTIMATION

In this section, different types of capacitance acting on the wire are explained in detail such as parallel plate and fringing field capacitance.

7.4.3.1 Parallel Plate Capacitor

We have already discussed the formation of capacitance between two parallel plates in Subsection 7.3.3. Considering the overall area of metal plate A and distance between plates as T_x, the parallel plate capacitance can be obtained as

$$C_{p2p} = \frac{\varepsilon_{r,die}}{T_x} A = \frac{\varepsilon_{r,die}}{T_x} W_x L_x \tag{7.57}$$

$$C_{p2p} = \frac{\varepsilon_r \times \varepsilon_0}{T_x} W_x L_x \tag{7.58}$$

7.4.3.2 Fringing Capacitance

As the CMOS technology continues shrinking and their feature size reducing, the thickness of wire is not scaled with the same factor. The reason behind not scaling the thickness is due to maintaining the overall cross-sectional area as large as possible by keeping the minimum sheet resistance [21]. Due to that, the capacitance between sidewall and substrate occurs significantly which is known as a fringing capacitance, as shown in Figure 7.5. In the deep submicron technology, the impact of fringing capacitance cannot be ignored. The fringing capacitance of a wire can be obtained as

$$C_{fringe} = 2\varepsilon_{r,die} \ln\left(1 + \frac{T_x}{h_g}\right) \tag{7.59}$$

where the digit 2 indicates that both sides of the wire have been considered.

The total capacitance in per unit length can be obtained by the sum of the parallel plate and fringing capacitance as

$$C_t = C_{p2p} + C_{fringe} \tag{7.60}$$

7.4.3.3 Lateral Capacitance

Apart from parallel and fringing capacitance, the lateral capacitance also has a major impact on interconnect and mainly formed due to two-wire placed closely to each other on the same plane. The lateral capacitance in *p.u.l.* for the closely spaced wires can be calculated as

$$C_{lateral} = \frac{\varepsilon_{r,die} T_x}{s} \tag{7.61}$$

Example 7.5:

Consider the wire made of metal 1 placed above the dielectric constant with a thickness of 0.62 μm. Calculate the total capacitance in *p.u.l.* and total capacitance for 100 μm length having a width of 0.6 μm and thickness of 0.7 μm, respectively.

SOLUTION:

For a given data, we can calculate the parallel plate capacitance as

$$C_{p2p} = \frac{\varepsilon_r \times \varepsilon_0}{T_x} W_x L_x$$

$$C_{p2p} = \frac{3.9 \times 8.854 \times 10^{-12}}{0.7 \times 10^{-06}} 0.6 \times 10^{-06}$$

$$C_{p2p} = 29.598 \text{ aF/µm}$$

Similarly, the fringing capacitance can be calculated as

$$C_{fringe} = 2\varepsilon_{r,die} \ln\left(1 + \frac{T_x}{h_g}\right)$$

$$C_{fringe} = 2 \times 3.9 \times 8.854 \times 10^{-12} \ln\left(1 + \frac{0.7 \times 10^{-06}}{0.62 \times 10^{-06}}\right)$$

$$C_{fringe} = 52.187 \text{ aF/µm}$$

The total capacitance of the wire is the sum of the parallel plate and fringing capacitance and can be expressed in *p.u.l.* as

$$C_t = C_{p2p} + C_{fringe}$$

$$C_t = 29.598 + 52.187 = 81.785 \text{ aF/µm}$$

At 100 µm of interconnect length, the total capacitance can be obtained as

$$C_{wire} = C_t \times L_x$$

$$C_{wire} = 81.785 \times 100 \times 10^{-06}$$

$$C_{wire} = 8.1785 \text{ fF}$$

Example 7.6:

Using 180 nm technology, assuming the following parameter based on the ITRS benchmark as width = 0.6 µm, thickness $T_x = 0.7$ µm, dielectric thickness $h_g = 0.62$ µm, and spacing between the wire $s = 0.6$ µm.

  a. Calculate parallel plate, lateral and fringing capacitance.
  b. Calculate middle wire capacitance when it is closely packed.
  c. Calculate middle wire capacitance when it is separated with wider space.

SOLUTION:

  a. We can obtain the parallel plate, lateral, and fringing capacitance using equations (7.58), (7.59), and (7.61) as

The parallel plate capacitance will be

$$C_{p2p} = \frac{\varepsilon_r \times \varepsilon_0}{T_x} W_x L_x$$

$$C_{p2p} = \frac{3.9 \times 8.854 \times 10^{-12}}{0.7 \times 10^{-06}} 0.6 \times 10^{-06}$$

$$C_{p2p} = 29.597 \text{ aF/µm}$$

And lateral capacitance

$$C_{lateral} = \frac{\varepsilon_{r,die} T_x}{S}$$

$$C_{lateral} = \frac{3.9 \times 8.854 \times 10^{-12} \times 0.7 \times 10^{-06}}{0.6 \times 10^{-06}}$$

$$C_{lateral} = 40.286 \text{ aF/µm}$$

The fringing capacitance of interconnect can be obtained as

$$C_{fringe} = 2\varepsilon_{r,die} \ln\left(1 + \frac{T_x}{h_g}\right)$$

$$C_{fringe} = 2 \times 3.9 \times 8.854 \times 10^{-12} \ln\left(1 + \frac{0.7 \times 10^{-06}}{0.62 \times 10^{-06}}\right)$$

$$C_{fringe} = 52.187 \text{ aF/µm}$$

b. The capacitance of the middle wire can be calculated when it is closely spaced

$$C_{wire} = 2C_{p2p} + 2C_{lateral} + C_{fringe}$$

$$C_{wire} = 2(29.597) + 2(40.286) + (0.00)$$

$$C_{wire} = 139.766 \text{ aF/µm}$$

c. The capacitance of the middle wire can be calculated when it is widely spaced

$$C_{wire} = 2C_{p2p} + 2C_{lateral} + C_{fringe}$$

$$C_{wire} = 2(29.597) + 2(0.00) + (52.187)$$

$$C_{wire} = 111.381 \text{ aF/µm}$$

7.5 CALCULATION OF INTERCONNECT DELAY

Since the last decade, the overall performance and robustness of interconnect have degraded with shrinking technology. It has resulted in significant attention toward the interconnect delay at the higher operating frequency. As per the current layout structure, it has been observed that interconnect has closer proximity and higher density that shows a significant impact on the overall performance of VLSI technology. In the early stage of the VLSI industry, only the impact of capacitance was considered, which was enough for the delay model due to lower operating speed. With the advancement of technology and complex geometry, for longer interconnect, the resistance has started dominating the overall performance. Therefore, for advanced technology, the impact of inductance has become an important factor that needs to be considered accurately while modeling. Currently, in most of the circuit, interconnect delay contributes almost 70% delay in comparison to the overall delay. Therefore, interconnect delay cannot be ignored while modeling the VLSI circuit. In order to understand the impact of parasitics (resistance, capacitance, and inductance), different delay model has been proposed to simplify the electrical model of interconnect.

7.5.1 *RC* Delay Model

The performance of the interconnect can be estimated using the *RC* delay model that can be categorized as a lumped and distributed model [22]. The most commonly used models are *RC*-lumped and *RC*-distributed model in the early stage of the VLSI industry.

1. ***RC*-lumped model:** For beginners, this model is the simplest electrical model of interconnect that consists of a single time constant of the *RC* circuit, as shown in Figure 7.11. When only one aspect of circuit behaviors is required to understand, then considering only a single *RC*-lumped circuit is enough while it has a small inductive effect and less operating frequency. The resistance and capacitance values exhibited by an ideal wire are zero. If we assume the interconnect exhibits resistance and capacitance as R_{int} and C_{int} for an interconnect length of L_x, we can estimate the parasitics of the interconnect as

$$R_w = R_{int} \times L_x \qquad (7.62)$$

$$C_w = C_{int} \times L_x \qquad (7.63)$$

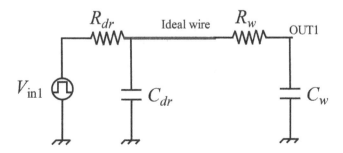

FIGURE 7.11   A simple lumped *RC* network.

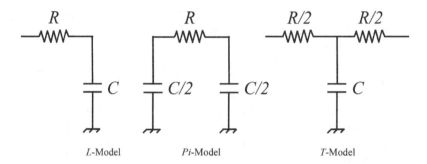

| L-Model | Pi-Model | T-Model |

FIGURE 7.12   Electrical equivalent models of interconnect.

where the *p.u.l.* resistance and capacitance of interconnect can be denoted as R_{int} and C_{int}, respectively. The time constant of the interconnect driven by RC driver can be computed as

$$\tau_{wire} = (R_{dr} \times C_{dr}) + (R_{dr} + R_w)C_w \tag{7.64}$$

The lumped model can be configured in three different ways: *L*-, *Pi*-, and *T*-model as, shown in Figure 7.12.

Assuming that the series RC electrical model of interconnect has a constant voltage supply at time $t = 0$, then the output voltage across the capacitor of the simplest RC model can be obtained as

$$V_{out}(t) = V_{in}\left(1 - e^{-\frac{t}{RC}}\right) \tag{7.65}$$

The 50% input–output delay for the rising output at $t = \tau_{50\%}$ can be computed as

$$V_{out}(t) \doteq \frac{V_{in}}{2} \tag{7.66}$$

$$\frac{V_{in}}{2} = V_{in}\left(1 - e^{-\frac{\tau_{50\%}}{RC}}\right) \tag{7.67}$$

$$\tau_{50\%} = RC\ln(1/0.5) = 0.6931RC \tag{7.68}$$

Hence, the 50% propagation delay for an RC network is

$$\tau_{50\%} = 0.6931RC$$

Similarly, 10% and 90% propagation delay can be found as

$$0.1V_{in} = V_{in}\left(1 - e^{-\frac{\tau_{10\%}}{RC}}\right)$$

$$\tau_{10\%} = RC\ln(1/0.9) = 0.1054RC \tag{7.69}$$

And for 90%, $0.9V_{in} = V_{in}\left(1 - e^{-\frac{\tau_{90\%}}{RC}}\right)$

$$\tau_{90\%} = RC\ln(1/0.1) = 2.3026RC \qquad (7.70)$$

Therefore, 10–90% propagation delay can be obtained as $\tau_{10-90\%} = \tau_{90\%} - \tau_{10\%}$

$$\tau_{10-90\%} = 2.3026RC - 0.1054RC$$

$$\tau_{10-90\%} = 2.2RC \qquad (7.71)$$

As we know, the accuracy of the lumped network is an approximation of the actual transient response of interconnect. The accuracy of the *L*-model can be significantly improved by splitting the resistance at both the ends as *T*-model, as shown in Figure 7.12.

2. **RC-distributed model:** Though the *RC*-lumped model is good for short wires, it is pessimistic and inaccurate for longer wires. Consequently, the distributed *RC* model is preferred for longer wires in which the resistance and capacitance of the wire are distributed along its entire length, as shown in Figure 7.13.

   Let the total interconnect resistance and capacitance be denoted as *R* and *C*, respectively along with the length l_x. The interconnect line is divided into *N*-segment having distributed resistance (*r*) and capacitance (*c*) of each line in per unit length. It can be defined as

$$r = \frac{R}{l_x}, \ c = \frac{C}{l_x}, \ and \ \Delta l_x = \frac{l_x}{N} \qquad (7.72)$$

$$\tau_{delay} = \sum_{i=1}^{N}\left(ir\Delta l_x\right)c\Delta l_x = rc\left(\Delta l_x\right)^2\left(1 + 2 + 3 + \ldots + N\right) = rc(l_x/N)^2\left(\frac{1+N}{2}\right)N \qquad (7.73)$$

$$= rcl_x^2\left(\frac{1+N}{2N}\right) \qquad (7.74)$$

$$\tau_{delay} = \lim_{N\to\infty} rcl_x^2\left(\frac{1+N}{2N}\right) = \frac{rcl_x^2}{2} = \frac{RC}{2} \qquad (7.75)$$

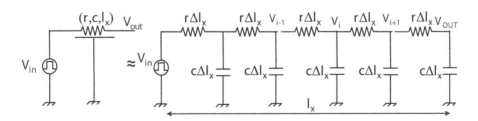

FIGURE 7.13 *N*-segment distributed *RC* network.

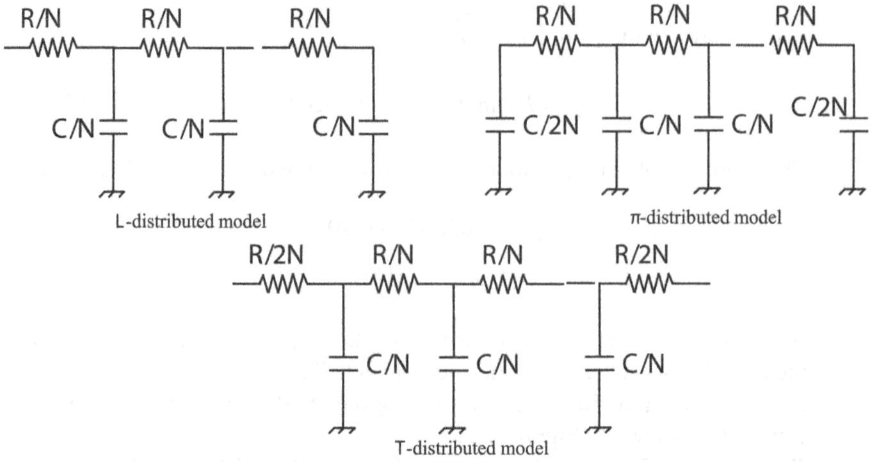

R/N   R/N   R/N   R/N   R/N   R/N

C/N   C/N   C/N   C/2N   C/N   C/N   C/2N

L-distributed model   π-distributed model

R/2N   R/N   R/N   R/2N

C/N   C/N   C/N

T-distributed model

FIGURE 7.14 The distribution of L, π, and T-model into N-segment.

Therefore, it can be observed that the delay of the wire is proportional to the square of its length. Further, the Elmore delay of distributed network exhibits approximately 2 times improved performance as observed from expressions (7.68) and (7.75).

Moreover, the lumped model, as shown in Figure 7.12 can also be represented in distributed form for L, π, and T model as depicted in Figure 7.14.

7.5.2 Elmore Delay Model

The Elmore delay has proven to be extremely useful formulas. Apart from investigating interconnect using the Elmore delay, it helps the reader to understand the impact of delay approximation by considering a simple RC network as described in **Section 7.3.4** [14, 23–25]. Further, for an N-stage RC cascade connection, as shown in Figure 7.15, the propagation delay can be obtained based on Elmore delay as

$$\tau_{Elmore} = NRC + (N-1)RC + \cdots + 2RC + RC$$
$$= \frac{1}{2}N(N+1)RC \propto N^2 RC$$

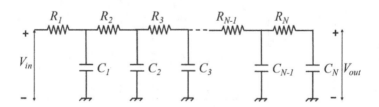

FIGURE 7.15 N-stage RC network in cascade form.

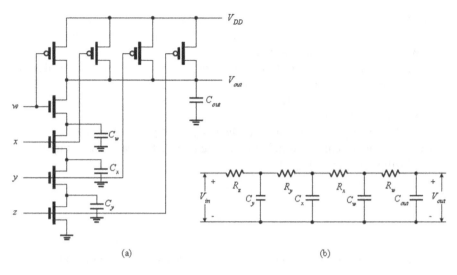

FIGURE 7.16 Four input NAND gate (a) logic circuit and (b) equivalent electrical network.

Example 7.7:

Find out the propagation delay of four input NAND gate, as shown in Figure 7.16 using Elmore delay.

SOLUTION:

Assuming that the output capacitance is the sum of all the capacitive parasitics of NMOS and PMOS transistors, the worst-case delay can be observed when the output voltage is discharged to the ground. In this case, the input voltage is zero. Hence, the propagation delay at the output node for an RC cascade network is

$$\tau_{Elmore(out)} = C_y R_z + C_x \left(R_y + R_z\right) + C_w \left(R_x + R_y + R_z\right) + C_{out} \left(R_w + R_x + R_y + R_z\right)$$

7.5.3 TRANSFER FUNCTION MODEL BASED ON ABCD PARAMETER MATRIX

The performance of the interconnect can be validated analytically using the ABCD parameter matrix. The ABCD parameters are constant and used to analyze the transmission line [26]. The input voltage and current supplied to passive, linear, and bilateral elements of the transmission line can be expressed in terms of output voltage and current. Considering two-port or four-terminal networks, the input voltage can be expressed in terms of output voltage and current as

$$V_{in} = AV_{out} + BI_{out} \tag{7.76}$$

$$I_{in} = CV_{out} + DI_{out} \tag{7.77}$$

where the circuit constant parameters of the transmission line are denoted by A, B, C, and D. These constant parameters values can be calculated using the above expression. We can consider the following two cases:

Case 1. Open a terminal of receiving end: In this case, the current flowing through the load will be zero, i.e., $I_{out} = 0$

$V_{in} = AV_{out}$ and $A = \frac{V_{in}}{V_{out}}$ unitless

From equation (7.77)

$I_{in} = CV_{out}$ and $C = \frac{I_{in}}{V_{out}}$ unit mho that shows open circuit conductance.

Case 2. Short circuit terminal of receiving end: In this case, the voltage across the load will be zero, i.e., $V_{out} = 0$

$$V_{in} = BI_{out}$$

$$B = \frac{V_{in}}{I_{out}}$$

The unit will be the same as impedance and denoted as ohm sometimes also called short circuit resistance.

Similarly, $I_{in} = DI_{out}$ and $D = \frac{I_{in}}{I_{out}}$ unitless

The $ABCD$ parameters have some unique features such as

(a) For a given system, it follows

$$A = D \text{ and } AD - BC = 1$$

7.5.4 FINITE DIFFERENCE TIME DOMAIN MODEL

It is one of the most powerful numerical techniques. Firstly, FDTD was developed to solve the time-domain Maxwell's equations [27–29]. Later on, it was adopted to solve the transmission line by segmenting the interconnect lines into multiple small segments in space and time, as shown in Figure 7.17. The interconnect line can be described by the telegrapher's equation whereas the associated parameters are voltage (V), current (I), as shown in Figure 7.18, and interconnect line is modeled by the resistance, capacitance, and inductance elements. After modeling the wire into the RLC element, the interconnect line is established within the boundary condition of the near and far end region. Using driver-interconnect-load setup, either a resistive driver or a CMOS driver can be used at the near end to drive the interconnect lines that are terminated with load capacitance at the far end region. After that, the boundary condition must be incorporated within the computation region to obtain the overall accurate performance. This method provides higher accuracy if it obeys the following

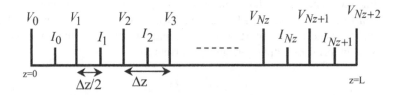

FIGURE 7.17   Distribution of voltages and current along interconnect of length L.

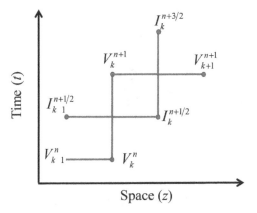

FIGURE 7.18 Voltage and current flow with time and space for interconnect line.

conditions such as (i) spatial step size must be small enough as compared to the wavelength, (ii) time step must be small enough to satisfy the Courant condition.

The capacitive coupled interconnect line can be described by the Telegrapher's equations as

$$\frac{d}{dz}V(z,t)+RI(z,t)+L\frac{d}{dt}I(z,t)=0 \tag{7.78a}$$

$$\frac{d}{dz}I(z,t)+GV(z,t)+C\frac{d}{dt}V(z,t)=0 \tag{7.78b}$$

where V and I are $M \times 1$ voltage and current column vector along the coupled transmission lines and can be observed as

$$V = \begin{bmatrix} V(1) \\ V(2) \\ V(3) \\ \vdots \\ V(M) \end{bmatrix} \text{ and } I = \begin{bmatrix} I(1) \\ I(2) \\ I(3) \\ \vdots \\ I(M) \end{bmatrix}, \text{ respectively}$$

The parasitics of interconnect such as resistance, inductance, conductance, and capacitance in $p.u.l.$ of the coupled line of $M \times M$ matrices can be expressed as

$$[R] = \begin{bmatrix} R_d^1 & 0 & 0 & 0 \dots & 0 \\ 0 & R_d^2 & 0 & 0 \dots & 0 \\ 0 & 0 & R_d^3 & 0 \dots & 0 \\ \vdots & 0 & 0 & 0 \dots & 0 \\ 0 & \dots & 0 & 0 \dots & R_d^N \end{bmatrix}_{N \times N}, [L] = \begin{bmatrix} L_d^1 & M_j^{(1,2)} & M_j^{(1,2)} & \dots & M_j^{(1,N)} \\ M_j^{(2,1)} & L_d^2 & M_j^{(2,3)} & \dots & M_j^{(2,N)} \\ M_j^{(3,1)} & M_j^{(3,2)} & L_d^3 & \dots & M_j^{(3,N)} \\ \vdots & \vdots & \vdots & \dots & \vdots \\ M_j^{(N,1)} & M_j^{(N,2)} & M_j^{(N,3)} & \dots & L_d^{(N)} \end{bmatrix}_{N \times N},$$

$$\text{and } [C] = \begin{bmatrix} C_d^1 + \displaystyle\sum_{i=2}^{N} C_c^{1,i} & -C_c^{1,2} & -C_c^{1,3} \cdots & -C_c^{1,N} \\[2ex] -C_c^{2,i} & C_d^2 + \displaystyle\sum_{i=2}^{N} C_c^{2,i} & -C_c^{2,3} \cdots & -C_c^{2,N} \\[2ex] \vdots & \vdots & \vdots \cdots & \vdots \\[2ex] 0 & 0 & 0 \;\; 0 & C_d^N + \displaystyle\sum_{i=2}^{N-1} C_c^{N,i} \end{bmatrix}_{N \times N}$$

whereas, R_d, C_d, C_c, are the distributed resistance, capacitance, coupling capacitance, respectively. The series inductance L_d^N is the sum of magnetic L_e^N and kinetic inductance L_k^N and can be denoted as $L_d^{(N)} = L_k^N + L_e^N$. The $M_j^{(N,1)}$ is referred to as mutual inductance and it can be considered zero if the impact is neglected.

Using the central difference method on discretized interconnect line into equations (7.78a) and (7.78b), the voltage and current expression can be obtained as

$$V_k^{n+1} = [E][F]V_k^n + \{[E]/\Delta z\}\left[I_{k-1}^{n+1/2} - I_k^{n+1/2}\right] \tag{7.79}$$

$$I_k^{n+3/2} = [B][D]I_k^{n+1/2} + [B]\left[V_k^{n+1} - I_{k+1}^{n+1}\right] \tag{7.80}$$

where $[E] = (C/\Delta t + G/2)^{-1}$, $[F] = (C/\Delta t - G/2)$, $[B] = \left[(\Delta z/\Delta t)L + (\Delta z/2)R\right]^{-1}$, and $[D] = \left[(\Delta z/\Delta t)L - (\Delta z/2)R\right]$

At the interconnect near end terminal, the current and voltage are represented by I_0 and V_0, respectively as

$$V_0^{n+1} = \left\{1/\left[(C_{tr}/\Delta t) + 1/R_{tr}\right]\right\}\left[(C_{tr}/\Delta t)V_0^n + (U/R_{tr})V_s^{n+1} - I_0^{n+1}\right] \tag{7.81}$$

$$I_0^{n+1} = \left(1/R_{fix}\right)\left[V_0^{n+1} - V_1^{n+1}\right] \tag{7.82}$$

The current and voltage at far end terminal of interconnect are represented as

$$V_{N_Z+1}^{n+1} = \frac{EF}{\left[U + \frac{E}{2\Delta z R_{fix}}\right]}V_{N_Z}^1 + \frac{E}{\Delta z\left[U + \frac{E}{2\Delta z R_{fix}}\right]}\left[\frac{V_{N_Z+2}^{n+1}}{2R_{fix}} + I_{N_Z}^{n+1/2} - \frac{1}{2}I_{N_Z+1}^n\right] \tag{7.83}$$

$$I_{N_Z+1}^{n+1} = \left(1/R_{fix}\right)\left[V_{N_Z+1}^{n+1} - V_{N_Z+2}^{n+1}\right] \tag{7.84}$$

$$V_{N_Z+2}^{n+1} = V_{N_Z+2}^n + \left(\Delta t/C_L\right)I_{N_Z+1}^n \tag{7.85}$$

The propagation delay can be obtained by solving the voltage and current expression (7.78) to (7.85) in space and time for interconnect.

7.6 ESTIMATION OF INTERCONNECT CROSSTALK NOISE

The interconnect performance mainly affected by the inductive and capacitive phenomenon generated between the coupled interconnect line [30, 31]. The crosstalk effect can degrade or is responsible for digital logic faults or data loss. The major source of crosstalk is coupling capacitance between the lines, as depicted in Figures 7.19 and 7.20.

The crosstalk in interconnects primarily depends on the switching characteristics of coupled lines. Depending on the switching behavior, crosstalk can be categorized into functional and dynamic types [32]. With reference to a ground line, when an aggressor line is excited with a pulse input, the victim line experiences a voltage spike as a peak noise known as the functional crosstalk, as shown in Figure 7.20(a). The VLSI designer needs to be careful while designing interconnect to have an acceptable range of extra peak noise such that it cannot give any false logic state on the victim line. On the other hand, dynamic crosstalk can be experienced when the aggressor and the victim line are excited by simultaneous pulse inputs [33, 34]. Under this category, the input switching for the aggressor and victim lines be either in the same direction (in-phase) or in the opposite direction (out-of-phase), as depicted in Figure 7.20(b). In order to have analytical exposure of peak noise V_{peak}, the model shown in Figure 7.19 can be derived using step-input pulse as

$$\frac{V_{peak}}{V_{dd}} = \frac{\left(R_{dr} + R_{intV}\right)C_c}{\left(R_{dr} + R_{intA}\right)\left(C_{intA} + C_{LA} + C_c\right) + \left(R_{dr} + R_{intV}\right)\left(C_{intV} + C_{LV} + C_c\right)} \quad (7.86)$$

where the peak noise and supply voltage are denoted as V_{peak} and V_{dd}. The R_{dr} and C_{dr}, R_{intA} and R_{intV} are the effective driver resistance and capacitance, lumped resistance

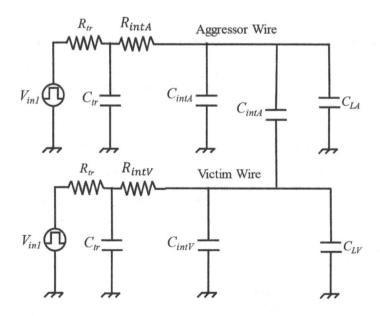

FIGURE 7.19 A capacitive crosstalk modeling of coupled interconnect.

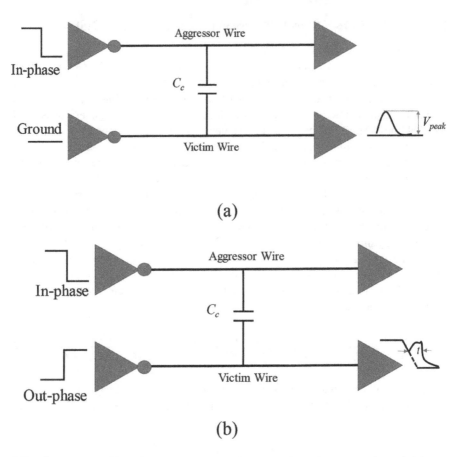

FIGURE 7.20 (a) Effect of crosstalk noise (b) 50% extra delay due to opposite switching.

of aggressor and victim wire, respectively. The C_{intA} and C_{intV}, C_{LA} and C_{LV} are the lumped capacitance of aggressor and victim wire, load capacitance of aggressor, and victim fanout gate. The coupling capacitance can be denoted as C_c.

7.7   ESTIMATION OF INTERCONNECT POWER DISSIPATION

As we know that the load capacitor (C_L) consumes energy when an input pulse is applied with amplitude to interconnect lines. The performance of the DIL system depends not only on delay but also on the power dissipation. Therefore, the power dissipation can be categorized as dynamic and static power dissipation. The dynamic power dissipation is due to the charging and discharging of load capacitance [35], whereas the static power dissipation is the leakage that occurs when the system is not powered or in standby mode.

   1. **In case of dynamic power dissipation:** When a certain amount of energy is drawn from the power supply, then the voltage rises from 0 to V_{dd} during the charging of load capacitance through PMOS transistor and the stored

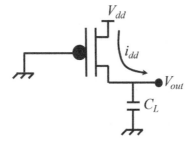

FIGURE 7.21 Equivalent electrical circuit during low to high switching input.

energy is charged in the PMOS transistor. On the other hand, the stored energy dissipates in the NMOS transistor due to the discharging of load capacitance during the high to low transition. Using an equivalent electrical network, as shown in Figure 7.21, we can derive an expression for energy consumption for low to high input switching while assuming that the input pulse has zero rise and fall time. During low to high switching, the energy E_{dd} required is drawn from the supply and at the end terminals, the capacitor is used to store the charged energy E_C. The corresponding output voltage and current supply can be observed in Figures 7.22(a) and 7.22(b) and the energy can be expressed as

$$E_{V_{dd}} = \int_0^\infty i_{V_{dd}}(t)V_{dd}dt = V_{dd}\int_0^\infty C_L \frac{dV_{out}}{dt}dt = C_L V_{dd}\int_0^{V_{dd}} dV_{out} = C_L V_{dd}^2 \qquad (7.87)$$

$$E_C = \int_0^\infty i_{V_{dd}}(t)V_{out}dt = C_L\int_0^\infty \frac{dV_{out}}{dt}V_{out}dt = C_L V_{out}\int_0^{V_{dd}} dV_{out} = \frac{C_L V_{out}^2}{2} \qquad (7.88)$$

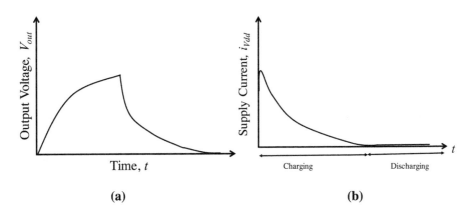

FIGURE 7.22 (a) Output voltage and (b) supply current during charging and discharging of load capacitance.

The power consumption is measured by observing how many times it is getting switched from low to high and high to low [35, 36]. Assuming that the gate is triggered with on and off $f_{0\to1}$ times each second, then the power consumption is given as

$$P_{dyn} = C_L V_{dd}^2 f_{0\to1} \qquad (7.89)$$

where the frequency of energy-consuming transitions is denoted as $f_{0\to1}$.

Example 7.8:

For CMOS inverter terminated with a load capacitance of 6 fF and driven by a supply voltage of 2.5 V.

a. Calculate the amount of energy needed to charge and discharge of load capacitance.
b. Assume that the CMOS inverter is switched at the maximum possible rate of $2t_p$, where $t_p = 32.5$ psec, then find the dynamic power dissipation of the circuit.

SOLUTION:

a. As we know, the energy required for load capacitance for dynamic switching is

$$E_{dyn} = C_L V_{dd}^2$$

$$E_{dyn} = (6fF)(2.5)^2 = 37.5 \text{ fJ}$$

b. The dynamic power dissipation can be calculated as

$$P_{dyn} = C_L V_{dd}^2 f_{0\to1} = C_L V_{dd}^2 \frac{1}{T} = C_L V_{dd}^2 \frac{1}{2t_p}$$

$$P_{dyn} = E_{dyn} \frac{1}{2t_p} = (34.5fJ)\left(\frac{1}{32.5 \text{ ps}}\right)$$

$$P_{dyn} = 580 \text{ } \mu W$$

2. **Static Power Dissipation:** The steady-state or static power dissipation of an electrical circuit can be expressed as

$$P_{stat} = I_{stat} V_{dd} \qquad (7.90)$$

In CMOS inverter, the PMOS and NMOS transistors are never acting simultaneously during a steady-state region and hence the static current flowing through the inverter is equal to zero. Unfortunately, as shown in

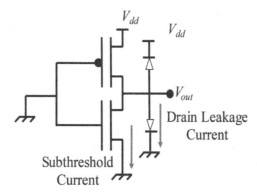

FIGURE 7.23 CMOS inverter represents the source of the leakage currents at zero input voltage.

Figure 7.23, the leakage current will flow in the reverse-biased junction of the diode, which is located between the source or drain and substrate. In addition to that, the leakage current can also be observed due to thermally generated carriers and that can be increased with an increase in the junction temperature of the transistor.

7.8 SUMMARY

This chapter can be summarized as follows:

- A rigorous and in-depth information starting from interconnect scaling issues from micron technology to deep submicron technology.
- The electrical modeling of interconnect was demonstrated that is used to calculate the performance in terms of propagation delay, crosstalk effect, and power dissipation using interconnect parasitics.
- Performance analysis was demonstrated based on several delay models that are analytically presented for better understanding of the readers.
- Finally, this chapter was concluded with the short, long, and MCQ-type question to check how effectively did the reader gain information from this chapter.

7.9 MULTIPLE-CHOICE QUESTIONS

1. Which technique is used to obtain the parasitics of interconnect such as resistance, capacitance, and inductance in VLSI design?
   a. Floor planning
   b. Extraction
   c. Placement and routing
   d. Testing

2. Identify the noise that degrades the integrity of signal due to vias, impedance mismatch, stubs, and other interconnection discontinuities.
   a. Reflection noise
   b. Power/ground noise
   c. Crosstalk noise
   d. All of the above

3. In CMOS inverter, when the signal changes its value from input to output,
find the propagation delay of a gate is the/as............transition delay time
for the signal.
   a. Lowest
   b. Average
   c. Highest
   d. None of the above

4. The power dissipation varies with the ....of the power supply voltage
for static CMOS gate.
   a. Fourth
   b. Square
   c. Cube
   d. 1/8th power

5. The speed of the CMOS logic gate mainly depends on which parameters?
   a. Gain factor of MOS
   b. Load capacitance
   c. Supply voltage
   d. All of the above

6. Under which category, the power is dissipated due to charging and discharging of load capacitance in the CMOS circuit?
   a. Dynamic dissipation
   b. Static dissipation
   c. Both a and b
   d. None of the above

7. In which factor, the overall delay of the electrical circuit depends in accor-
dance with the scaling technology?
   a. The voltage through which capacitance must be charged
   b. Available current
   c. The capacitor to be charged
   d. All of the above

8. The occurrence of 'delay faults' depends on which factor?
   a. Aging effects & opens in metal lines connecting parallel transistors
   b. Improper estimation of on-chip interconnect & routing delays

    c. Variations in circuit delays & clock skews

    d. All of the above

9. The interconnect spacing is scaled by which factor?

    a. α

    b. $1/\alpha^2$

    c. α^2

    d. $1/\alpha$

10. Which impact can be observed in the die size due to a decrease in device dimension?

    a. Increases

    b. Decreases

    c. Does not affect

    d. Decreases and then increases

11. The crosstalk noise..........during the increase in operating frequency

    a. Increases

    b. Decreases

    c. Does not change

    d. Doubles

12. Identify the factors that define the size of a transistor

    a. Channel length

    b. Feature size

    c. Width

    d. Thickness "d"

13. Which factor is mostly considered for scaling?

    a. Constant electric scaling

    b. Constant voltage scaling

    c. Constant electric and voltage scaling

    d. Constant current model

14. Which factor is responsible for the wiring capacitance?

    a. Fringing fields

    b. Interlayer capacitance

    c. Peripheral capacitance

    d. All of the mentioned

15. Total wire capacitance is equal to

    a. Area capacitance

    b. Fringing field capacitance

    c. Area capacitance + fringing field capacitance

    d. Peripheral capacitance

7.10   SHORT ANSWER QUESTIONS

1. List out three major issues of scaling the Cu interconnect.
2. How the resistance of interconnect gets affected due to scaling?
3. Define the rise time and fall time of input waveform.
4. Define fringing capacitance and area capacitance.
5. Explain the reason behind choosing a polysilicon layer for local interconnect.
6. What is the pitch and aspect ratio of interconnect?
7. Define sheet resistance of interconnect.
8. Define lateral capacitance.
9. Differentiate between the lumped and distributed RC models.
10. Explain any two methods to reduce the impact of crosstalk effects on interconnect.

7.11   LONG ANSWER QUESTIONS

1. Explain the interconnect problem in Al and Cu material.
2. Explain the different scaling categories of electrical modeling.
3. Explain in detail
    i.   Resistive interconnect
    ii. Capacitive interconnect
    iii. Inductive interconnect
4. Derive an expression for the series RC circuit using ABCD parameters.
5. Derive an expression for the RLC circuit based on ABCD parameters.
6. Explain the crosstalk effect with a neat and clean diagram.
7. Explain the types of power consumption in interconnect.
8. Explain in which situation the impact of inductance can be ignored.
9. Consider a rectangular interconnect structure that exhibits a resistance of 69 mΩ/□, capacitance of 0.19 fF/μm, and inductance of 0.75 pH/μm in per unit length, respectively.
    Calculate the range of wire length in which the inductance needs to consider if
    i.   The signal rise time is 4 ns.
    ii. The signal rise time of 36 ps.
10. Consider the wire made of metal 1 placed above the dielectric constant with a thickness of 0.85 μm. Calculate the total capacitance per unit length and total capacitance for 500 μm length having a width of 0.79 μm and thickness of 0.98 μm, respectively.

REFERENCES

1. Moore, G. E. 1965. Cramming more components onto integrated circuits. *Electronics* 38, no. 8: 114–17.
2. Cavin, R. V., Zhirnov, V. V., Herr, D. J. C., Avila, A. and Hutchby, J. 2006. Research directions and challenges in nanoelectronics. *Journal of Nanoparticle Research* 8, no. 6:841–858.

3. Wen, W., Brongersma, S. H., Hove, M. V. and Maex, K. 2004. Influence of surface and grain-boundary scattering on the resistivity of copper in reduced dimensions. *Applied Physics Letter* 84, no. 15: 2838–2840.

4. Banerjee, K. and Mehrotra, A. 2001. Global (interconnect) warming. *IEEE Circuits and Devices Magazine* 17, no. 5: 16–32.

5. Cheng, D. K. 1989. *Field and Wave Electromagnetics* (2nd edition), Addison-Wesley, Reading, MA.

6. Havemann, R. H. and Hutchby, J. A. 2001. High-performance interconnects: An integration overview. *Proceedings of the IEEE* 89, no. 5: 586–601.

7. Bohr, M. T. 1995. Interconnect scaling—The real limiter to high performance ULSI. *IEDM*, pp. 241–244.

8. Dally, W. J. and Towles, B. 2004. *Principles and Practices of Interconnection Networks*, Morgan Kaufmann, Boston, MA.

9. Cheng, C. K., Lillis, J., Lin, S. and Chang, N. H. 2000. *Interconnect Analysis and Synthesis*, John Wiley Press, New York, NY.

10. Gauthier, C. and Amick, B. 2002. Inductance: Implications and solutions for high-speed digital circuits: The chip electrical interface. *Proceeding of IEEE International Solid-State Circuits Conference*, 2: 565–565.

11. Yuan, T., Buchanan, D. A., Wei, C., Frank, D. J., Ismail, K. E., Shih-Hsien, L., Sai-Halasz, G. A., Viswanathan, R. G., Wann, H. J. C., Wind, S. J. and Hon-Sum, W. 1997. CMOS scaling into the nanometer regime. *Proceedings of the IEEE* 85: 486–504.

12. Rabaey, J. M., Chandrakasan, A. and Nikolic, B. 2009. *Digital integrated circuits—A design perspective* (2nd edition), PHI Learning.

13. Barke, E. 1988. Line-to-ground capacitance calculation for VLSI: A comparison. *IEEE Transactions on Computer-Aided Design* 7, no. 2: 295–298.

14. Abou-Seido, A. I., Nowak, B. and Chu, C. 2004. Fitted Elmore delay: A simple and accurate interconnect delay model. *IEEE Transactions on VLSI* 12, no. 7: 691–696.

15. Adler, V. and Friedman, E. G. 1998. Repeater design to reduce delay and power in resistive interconnect. *IEEE Transactions on Circuits and Systems II: Analog and Digital Signal Processing* 45, no. 5: 607–616.

16. Duato, J., Yalamanchili, S. and Ni, L. 2003. *Interconnection Networks: An Engineering Approach*, Morgan Kaufmann, Boston, MA.

17. International Technology Roadmap for Semiconductors (ITRS). [Online], http://www. itrs2.net.

18. Hanchate, N. and Ranganathan, N. 2006. A linear time algorithm for wire sizing with simultaneous optimization of interconnect delay and crosstalk noise. In *Proceedings of the 19th International Conference on VLSI Design*, pp. 283–290.

19. Ismail, Y. I. and Friedman, E. G. 2000. Effects of inductance on the propagation delay and repeater insertion in VLSI circuits. *IEEE Transactions on Very Large Scale Integration (VLSI) Systems* 82, no. 2: 195–206.

20. Gauthier, C. and Amick, B. 2002. Inductance: Implications and solutions for high-speed digital circuits: The chip electrical interface. *Proceeding of IEEE International Solid-State Circuits Conference* 2: 565–565.

21. Stellari, F. and Lacaita, A. L. 2000. New formulas of interconnect capacitances based on results of conformal mapping method. *IEEE Transactions on Electron Devices* 47, no. 1:222–231.

22. Boese, K. D., Kahng, A. B., McCoy, B. A. and Robins, G. 1993. Fidelity and near-optimality of Elmore-based routing constructions. In *Proceedings of I.E. International Conference on Computer Design (ICCD '93)*, pp. 81–84.

23. Sakurai, T. 1993. Closed-form expressions for interconnection delay, coupling, and crosstalk in VLSI's. *IEEE Transactions on Electron Devices* 40, no. 1: 118–124.

24. Chen, C. P., Chen, Y. P. and Wong, D. F. 1996. Optimal wire-sizing formula under the Elmore delay model. *Proceedings of the 33rd annual Design Automation Conference* ACM, pp. 487–490.

25. Fishburn, J. P. and Schevon, C. A. 1995. Shaping a distributed-RC line to minimize Elmore delay. *IEEE Transactions on Circuits and Systems I: Fundamental Theory and Applications* 42, no. 12: 1020–1022.

26. Sahoo, M. and Rahaman, H. 2014. An ABCD parameter based modeling and analysis of crosstalk induced effects in Multilayer Graphene Nano Ribbon interconnects. *IEEE International Symposium on Circuits and Systems (ISCAS)*, Melbourne VIC, pp. 1138–1142.

27. Cherry, P. C. and Iskander, M. F. 1995. FDTD analysis of high frequency electronic interconnection effects. *IEEE Transactions on Microwave Theory and Techniques* 43, no. 10: 2445–2451.

28. May, M. P., Taflove, A. and Baron, J. 1994. FDTD modeling of digital signal propagation in 3-D circuits with passive and active loads. *IEEE Transactions on Microwave Theory Technology* 42, no. 8: 1514–1523.

29. Zhang, X. and Mei, K. K. 1988. Time-domain finite difference approach to the calculation of the frequency dependent characteristics of microstrip discontinuities. *IEEE Transactions on Microwave Theory Technology* 36, no. 12: 1775–1787.

30. Kahng, A. B., Muddu, S. and Sarto, E. 2000. On switch factor based analysis of coupled RC interconnects. In *Proceedings of the 37th Annual Design Automation Conference*, pp. 79–84.

31. Vittal, A. and Sadowska, M. M. 1997. Crosstalk reduction for VLSI. *IEEE Transactions on Computer-Aided Design of Integrated Circuits and Systems* 16, no. 3: 290–298.

32. Vittal, A., Chen, L. H., Sadowska, M. M., Wang, K. P. and Yang, S. 1999. Crosstalk in VLSI interconnections. *IEEE Transaction on Computer Aided Design of Integrated Circuits and Systems* 18, no. 12: 1817–1824.

33. Majumder, M. K., Kaushik, B. K. and Manhas, S. K. 2014. Analysis of delay and dynamic crosstalk in bundled carbon nanotube interconnects. *IEEE Transactions on Electromagnetic Compatibility* 56, no. 6: 1666–1673.

34. Majumder, M. K., Das, P. K. and Kaushik, B. K. 2014. Delay and crosstalk reliability issues in mixed MWCNT bundle interconnects. *Microelectronics Reliability* 54, no. 11: 2570–2577.

35. Kapur, P., Chandra, G. and Saraswat, K. C. 2002. Power estimation in global interconnects and its reduction using a novel repeater optimization methodology. *Proceedings of the 39th Annual Design Automation Conference, ACM*.

36. Magen, N., Kolodny, A., Weiser, U. and Shamir, N. 2004. Interconnect power dissipation in a microprocessor. International Workshop on System Level Interconnect Prediction, pp. 7–13.

8 VLSI Design and Testability

8.1 PREAMBLE

The increased complexity of integrated circuits (ICs) may cause the degradation of performance due to defects and faults. In order to have smooth functionality and defect-free circuit and devices, the testing and observability of ICs need to be grown for deep submicron technology. Technology enhancement and higher complexity of IC become the bottleneck for the circuit to work correctly and difficult to test due to the faster clock speed. Therefore, the demand of delay testing and timing analysis are tremendously increased due to the higher operating frequency and miniaturization of technology. In general, the testing of circuit and devices can be done based on the three ways such as verification testing, manufacturing testing, and acceptance testing. The correctness of the circuits and devices can be done with the help of verification testing. The logic faults, parametric faults, and physical faults during the manufacturing of an IC can be done using the manufacturing test. The quality of the product and devices supplied by the VLSI industry need to be verified and accepted by the user only after completing the acceptance test or performing incoming inspection. Recently, several works consist of fault simulation, test generation, design methodology, and synthesis for testability. Therefore, this chapter provides a detailed information and understanding regarding the testing, observability, and various fault models.

8.2 BASIC DIGITAL TROUBLESHOOT

In order to check the performance and correct functionality of fabricated chips, first, it has to go through the testing process done within the lab environment. The lab environment is carefully arranged based on the requirements that should have the following features [1]:

1. Power dissipation by the whole circuit needs to be measured by varying the supply voltage, V_{DD}.
2. All the analog and digital input and output pins are examined for any malfunctions by providing real-world signal.
3. Stability of an input clock signal is required to monitor.
4. Slow and fast data transfer and exchange through the PCI is required to examine.

8.2.1 Manufacturing Test

The manufacturing testing is established to verify the operation of whole gates that are situated within the IC to observe the proper functionality of a chip. This process of manufacturing testing is very important due to the malfunction and defects occur

237

during the chip fabrications or accelerated life testing (i.e., during these process, chips are passed through high stress under high voltages and temperature operation). The defects that can be observed on a chip are metal-to-metal layers, missing or damaged Vertical Interconnect Access (VIA), mixing of thin oxide layer with the substrate or well, and discontinuous wires [1]. The integrity of input–output pins is also tested just after the verification of internal gates in a chip to observe any malfunctions.

8.2.2 TESTER AND TEST FIXTURES

A chip or a system is going through a series of stimuli that can be done with the help of a device, known as tester [1]. Further, the results, obtained with the help of test equipment, as depicted in Figure 8.1, are monitored and stored based on the operations [2].

There are three types of test fixtures required in order to test a chip. They are:

1. Testing of the unpacked die or wafer level can be done with the help of a probe card using a chip tester.
2. A package part is also tested that requires a load board.
3. Further, a bench level of testing can be performed using a Printed Circuit Board (PCB) that may be done with or without the help of tester.

8.2.3 TEST PROGRAMS

A high-level language is required to write the program in order to run the tester device for testing and verification operation. Moreover, the test program contains a set of input variables and a set of output variables based on the operation and

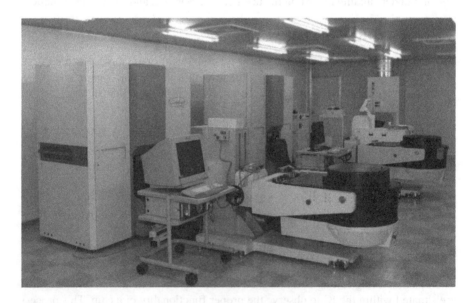

FIGURE 8.1   Test equipment used in the VLSI industry.

verification task [3]. The testing can be performed based on a comparative analysis between the asserted value and the final outcome of the test program. If the output of the test program does not match its reports, an error report is generated by setting the pattern into the test program. The set of variables that needs to be applied to the testers are:

- Supply voltage and clock are required to set,
- Physical tester pin and stimulus signal file must be mapped and assigned,
- The input and output pins of the tester are assigned along with their V_{OH}/V_{IL} levels.

8.3   EFFECT OF PHYSICAL FAULTS ON CIRCUIT BEHAVIOR

In this section, various fault models are briefly explained in order to understand the behavior of the circuits. The physical faults include defects in silicon substrate, oxide defects, process variation and environmental defects, contaminant and scratches defects, and photolithography defects [4]. These physical defects can introduce several logical and electrical circuits' faults. Therefore, the fault methods are helpful in order to check the reliability of the parts whether it is in a good or bad condition.

8.3.1   FAULT MODELS

Manufacturing of circuits and devices has to go through several testing procedures with the help of various fault models due to stressful operating condition [5]. With the help of fault models, the input test vectors are applied to Device Under Test (DUT) or Circuit Under Test (CUT) and compared it with the golden device/circuit. The faults can occur due to either physical and manufacturing defects or electrical faults that are described below.

8.3.1.1   Line Stuck-at Faults

When a particular line is stuck at either logic 1 or logic zero, it will remain at the same logic in the electrical circuit that will reduce the functionality of the circuit and provide an error log, which is known as Line stuck-at fault [6].

8.3.1.2   Transistor Stuck-at Faults

If the operation of single or more than one transistors is stuck either at ON state (during which the transistor will always have drain and gate shorted and it results in conduction mode) or OFF state (during which the transistor will always be in the nonconducting condition), then this phenomenon is known as transistor stuck-at faults.

8.3.1.3   Floating Line Faults

These kinds of faults are generally observed when one or more nodes in a circuit do not have a physical connection and remain isolated that means there are physical breakdown that creates disconnection of nodes in a circuit [7].

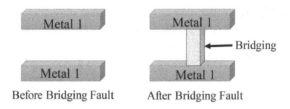

FIGURE 8.2   Fault observation in metal 1 wire.

8.3.1.4   Bridging Faults

These types of faults primarily occur due to excess of metal material or conducting materials between two or more lines or circuit nodes that are shorted and malfunctioned as depicted in Figure 8.2. These faults may also be generated due to the absence of an insulator between the lines [8].

8.4   TEST PRINCIPLES OF MANUFACTURING

In this section, first, the importance of different test methodologies for observing and controlling the behavior of an IC and systems is described. Later, the fault coverage is explained by incorporating different test vectors on circuit nodes. Further, Automatic test pattern generation approach is adopted to reduce the burden faced by the designer along with the delay fault testing in on-chip and off-chip circuit and systems [9].

8.4.1   Observability

In electronic circuits, the observability is the ability to measure the signal value at any node within the circuits while controlling the input of the circuit and observing its output. The observability plays an important role to check the particular node working as desired or not based on the operation [10]. From a good designer's point of view, it is required to have better observability to identify the output gate nodes. This technique should be adopted by the designer based on some basic design for test technique.

8.4.2   Controllability

Controllability is the process wherein a particular node of an internal circuit is either set to logic 1 or logic 0. This process is important in keeping a higher degree of testing environment for a particular signal within a circuit in an IC. The input pad is used to directly control the particular node. This phenomenon is quite difficult as it has to go through several test cycles to achieve the right state. Moreover, the designer has faced a challenge in providing an easy way to generate the test sequence to set the right state for poorly controlled node while considering the simple design for test methodology [11]. However, a good controllability can be set via a global reset signal by making all flip-flop resettable globally.

8.4.3 Fault Coverage

Fault coverage can be defined as the higher percentage of testing and observability of circuit within the chips that contains several nodes. Hence, for a good quality, the circuits are required to have 98.5% of fault coverage such that it can have access to most of the nodes within the IC. The fault coverage can be used to identify the faults by following the different aspects of fault detection techniques. The nodes of the circuit are kept on the test using either stuck the node at 1 or at 0. Further, each circuit within the chips is simulated with the test sequence and vectors and compared it with the golden circuit or golden machine [12]. The identification of fault can be observed if the faulty machine or circuit has a discrepancy with the golden machine. Figure 8.3 demonstrates a general block diagram wherein a test sequence is applied to digital circuits that are under test and generated output response. Further, the output response is compared with the stored input response to identify the faults.

8.4.4 Automatic Test Pattern Generation (ATPG)

The ATPG is adopted by most digital designers to reduce the complexity faced by the test engineer. During the past decades, the test engineering begged the circuit designer to incorporate some extra test circuits to check the functionality in an easy manner. As the technology advances, the speed of an IC is increased by incorporating more number of transistors in a chip. Therefore, the designer and manager have reduced the test circuitry to maintain the cost of a die such that more number of gates can be implemented within the chip. With the increased complexity and density of an IC, the burden of test engineering increases before the invention of ATPG [13]. After the commercial use of ATPG tool, that provides a huge relief to the test designer, it has been adopted by the digital designer to achieve excellent fault coverage.

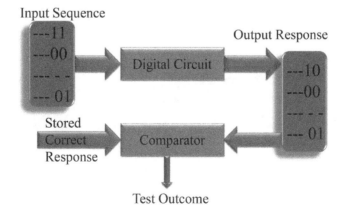

FIGURE 8.3   A general block schematic of fault detection.

8.4.5 Delay Fault Testing

Delay fault within a circuit can be understood with the help of CMOS inverter, where the higher delay can be observed if the circuit will not be functioned as implemented. This phenomenon can be observed if one of the PMOS or NMOS becomes open that introduces a higher delay and affect the timing. Moreover, the crosstalk is also one of the major causes of increased delay within the circuits [6]. Several software have already been modeled in order to test the delay and become an important factor as technology advanced and density increases that can produce a higher degree of failure. Therefore, monitoring of delay fault model is essential.

8.5 TEST APPROACHES

In order to improve the controllability and observability of the internal nodes of chips, the designer needs to integrate the design and test circuits to have a better Design For Testability (DFT) [11]. In order to have lower manufacturing cost for testing, the design should have a good observability and controllability that allows a high degree of fault coverage using a few sets of test vectors. This section demonstrates three main approaches that can be classified as ad hoc testing, scan design test, and built-in self-test (BIST).

8.5.1 Ad Hoc DFT Techniques

A tricky way to test the functionality of the digital circuit is adopted in order to reduce the complex and time-consuming systematic approach; the designer adopted this approach from several years termed as ad hoc testing [1, 7]. This technique follows some important points that are described below:

i. In order to reduce the cost of testing, the large circuits are segregated into several small sub-circuits. Testable chips first need to be divided at every level and for every functional block from architecture to circuit using multiplexer or scan chains, as shown in Figure 8.4.

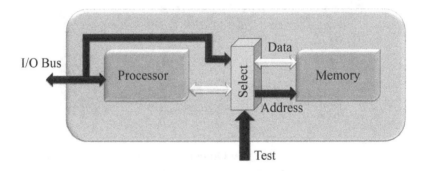

FIGURE 8.4   A multiplexer inserted between the blocks to control the selection.

ii. For better and improved controllability and observability, each test chip must be inserted with a test access point such that it can easily be available to test the chips.
iii. To enhance the predictability or functionality, the circuit (flip-flops) must be easily accessible and the reset pins need to be controllable from primary inputs to have more effective utilizations.
iv. Physical isolation should be provided between analog and digital circuitry because the testing procedure of analog and digital circuits are completely different due to the requirement of diverse equipment.

8.5.2 Scan Design Test

In VLSI industry, the circuits are designed based on the predefined standard design rules along with more than one Test Control (TC) pins that are required at the primary input. In order to test a chip, a shift register is used that can be obtained by replacing the flip-flop with Scanned Flip-Flop (SFF) connected in cascade form. The output of the first SFF will act as input for the neighbor blocks by creating a chain series. The first SFF block is straightforward by connecting with the input pin, known as SCAN-IN; similarly, that last block of SFF is connected with the output pin, termed as SCAN-OUT. The combinational logic of ATPG block is depicted in Figure 8.5 that is used to obtain the tests to identify all possible faults present [7]. Moreover, the manufacturing test utilizes the scan sequences that are obtained by converting ATPG test and the shift register test is also applied to perform the manufacturing test.

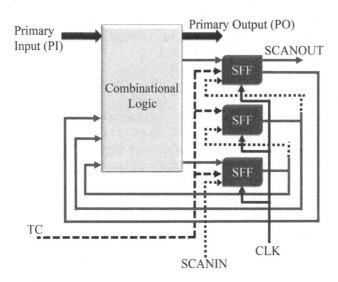

FIGURE 8.5 A block diagram of scan structure to a design.

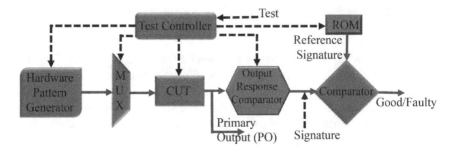

FIGURE 8.6   A block diagram of a typical BIST architecture with primary output.

8.5.3   BUILT-IN SELF-TEST (BIST)

Figure 8.6 presents the basic architecture of BIST that consists of three major hardware blocks such as a response analyzer, a test pattern generator, and a test controller [14]. The first step in this method is to generate the test pattern for the circuit that needs to keep into testing purposes. The generation of patterns can be done with the help of a counter, Read-Only Memory (ROM) with the stored pattern, and a Linear Feedback Shift Register (LFSR). After that, the test pattern applied to the test circuit is compared with the stored test pattern. Later, the signature analyzer can be done with the help of LFSR, where it analyzes the correctness of the test circuit using test response obtained during the process. The whole process can be done by activating the test using a control block such that the analysis can be smoothly done [15–16]. However, a test controller circuit is mostly used to perform several test-related functions. Moreover, the BIST demonstrates several benefits such as

1. The testing and maintenance cost are reduced due to reduced ATPG.
2. The storage and maintenance of test patterns are reduced.
3. The parallel process can be done for testing the circuits and reduces the time required to obtain the test results.

8.5.4   IDDQ TESTING

The presence of manufacturing fault can also be identified in CMOS integrated circuit using IDDQ testing. This technique allows the current to flow such that the fault can easily be identified in the quiescent state. Figure 8.7 shows a general block diagram of IDDQ testing, where the detection of fault can be observed by monitoring the IDDQ current. In order to have a faster and accurate analysis, the IDDQ test should be very sensitive [17].

8.6   DESIGN FOR MANUFACTURABILITY (DFM)

In engineering, the designers are used to design the product multiple times to achieve an effective way to make manufacturing the products easier and simpler. This can be done by optimizing the several manufacturing process such as

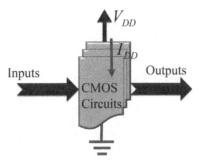

FIGURE 8.7 A block diagram of IDDQ testing.

fabrication, shipping, repair, delivery, procurement, assembly, and test [1, 7]. The demand of manufacturability can be increased by optimizing the circuits in the following ways:

1. **Physical:** The circuit designer needs to reduce the impact of process defects in order to increase the manufacturability at the physical level.
2. **Manufacturing Process:** In order to manufacture, the product with an appropriate process and technique can enhance the productivity and reliability.
3. **Redundancy:** The architecture is required to embed with the redundant structure in order to compensate for the defective components and elements on a chip. For example: if there is any defective word while testing the memory in the particular array, then it is reconfigured to a reductant array of extra row instead of using the defective array.
4. **Power:** The migration of metal can cause several failures in the circuits due to a rise in the power. The rise in the power can be observed due to an increase in current flow in the wire. Moreover, the RF devices or the devices that required higher power may cause performance degradation due to an increase in temperature over time. In order to overcome this issue, a suitable heat sinker or cooling fan should be used to remove the excess heat.

8.7   SYSTEM ON CHIP (SOC) TESTING

Testing of a system on a chip is known as SOC test, which is a challenging task due to complex circuits and devices. The design of SOC can be understood from placing multiple blocks on a single substrate like multiple devices on a single chip as depicted in Figure 8.8, and testing of such a system is quiet difficult. In order to have a feasible testing procedure, the designer has preinstalled the specialized and configurable embedded system on SOC such that the error and fault can easily be identified and rectified [7]. The use of an embedded system provides several advantages such as higher fault coverage, specific test speed, dynamic diagnostic options, and generation of test sequence for any random logic blocks.

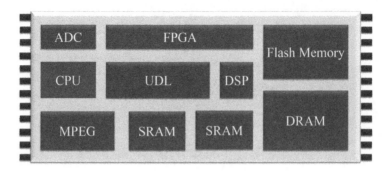

FIGURE 8.8 A design of SOC having multiple devices.

8.8 SUMMARY

The chapter summarizes the important factor that needs to be considered for easy and effective testability and observability of circuits, devices, and systems.

- A brief preamble was provided such that the reader can obtain enough information related on testing of a complex circuit by generating the test vectors.
- Several fault models were also demonstrated in order to identify the faults during the fabrication and under high stress due to high operating frequency.
- Finally, this chapter concluded with short, long, and MCQ-type questions to check the reader to effectively gain the information from this chapter.

8.9 MULTIPLE-CHOICE QUESTIONS

1. DFT is considered in the production of chips because:
   a. Manufactured chips are faulty and are required to be tested.
   b. The design of chips are required to be tested.
   c. Many chips are required to be tested within short interval of time, which yields timely delivery for the customers.
   d. All of the mentioned

2. ATPG stands for:
   a. Attenuated Transverse wave Pattern Generation
   b. Automatic Test Pattern Generator
   c. Aligned Test Parity Generator
   d. None of the mentioned

3. A metallic blob present between drain and the ground of the n-MOSFET inverter acts as:
   a. Physical defect
   b. Logical fault as output is stuck on 0
   c. Electrical fault as resistor short
   d. All of the mentioned

4. The functions performed during chip testing are:
   a. Detect faults in fabrication
   b. Detect faults in design
   c. Failures in functionality
   d. All of the mentioned

5. Delay fault is considered as:
   a. Electrical fault
   b. Logical fault
   c. Physical defect
   d. None of the mentioned

6. The fault simulation detects faults by:
   a. Test generation
   b. Construction of fault Dictionaries
   c. Design analysis under faults
   d. All of the mentioned

7. High resistance short present between drain and ground of n-MOSFET inverter acts as:
   a. Pull up delay error
   b. Logical fault as output is stuck at 1
   c. Electrical fault as transistor stuck on
   d. All of the mentioned

8. The ease with which the controller establish specific signal value at each node by setting input values is known as:
   a. Testability
   b. Observability
   c. Controllability
   d. Manufacturability

9. The poor controllability circuits are:
   a. Decoders
   b. Clock generators
   c. Circuits with feedback
   d. All of the mentioned

10. Large number of input vectors are used to set a particular node (1) or (0), to propagate an error at the node to output makes the circuit low on:
    a. Testability
    b. Observability
    c. Controllability
    d. All of the mentioned

11. Divide-and-conquer approach to large and complex circuits for testing is found in:
    a. Partition and Mux Technique
    b. Simplified automatic test pattern generation technique
    c. Scan based technique
    d. All of the mentioned

12. The ease with which the controller determines signal value at any node by setting input values is known as:
    a. Testability
    b. Observability
    c. Controllability
    d. Manufacturability

13. The circuits with poor observability are:
    a. ROM
    b. PLA
    c. Sequential circuits with long feedback loops
    d. All of the mentioned

14. Electromigration is a:
    a. Processing fault
    b. Material defects
    c. Time-dependent failure
    d. Packaging fault

15. IDDQ fault occurs when there is:
    a. Increased voltage
    b. Increased quiescent current
    c. Increased power supply
    d. Increased discharge

16. The quality of the test set is measured by:
    a. Fault margin
    b. Fault detection
    c. Fault correction
    d. Fault coverage

8.10   SHORT ANSWER QUESTIONS

1. Give some examples of practical BIST application in industry.
2. What are three types of testing in VLSI?
3. Define IDDQ testing.
4. Define bridge fault.
5. Define stuck open fault.
6. Define delay fault.

7. Define line delay fault.
8. What is the meaning of fault model?
9. Define controllability.
10. Define test, fault detection, and fault correction.

8.11 LONG ANSWER QUESTIONS

1. What is Built-in Self-Test? Discuss the issues and benefits of BIST. Describe BIST architecture and its operation.
2. Excluding the circuit under test, what are the four basic components of BIST and what function does each component perform?
3. Explain the logic circuit showing stuck at 1 fault and stuck at 0 faults. Also, explain how they are indistinguishable.
4. Explain the controllability and observability.
5. Differentiate between gate delay fault and transition fault.
6. Write a short note on Ad hoc testing.
7. Explain DFM.
8. Explain SOC testing.
9. Write a short note on automatic test pattern generation.
10. Explain any two methods of design testability.

REFERENCES

1. Abramovici, M., Breuer, M. A. and Friedman, A. D. 1996. *Digital Systems Testing and Testable Design* (2nd edition), IEEE Press, Piscataway.
2. Weste, N. H. E. and Eshraghian, K. 2004. *Principles of CMOS VLSI Design* (2nd edition). Addison-Wesley Publishing Co, Reading, MA, pp. 576.
3. Wadsack, R. L. 1978. Fault modeling and logic simulation of CMOS and MOS integrated circuits. *Bell System Technical Journal* 57, no. 5: 1449–1474.
4. Mokhari-Bolhassan, M. E. and Kang, S. M. 1988. Analysis and correction of VLSI delay measurement errors due to transmission-line effects. *IEEE Transactions on Circuits and Systems* 35, 19–25.
5. Dekker, R., Beenker, F. and Thijssen, L. 1990. A realistic fault model and test algorithms for static random access memories. *IEEE Transactions on Computer-Aided Design* 9, no. 6: 567–572.
6. Jha, N. and Kundu, S. 1990. *Testing and Reliable Design of CMOS Circuits*, Kluwer Academic Publishers, Norwell, MA.
7. Weste, N. H. E. and Eshraghian, K. 1993. *Principles of CMOS VLSI Design* (2nd edition), Addison-Wesley Publishing Co, Reading, MA.
8. Acken, J. M. 1983. Testing for bridging faults (shorts) in CMOS circuits. In *Proceedings of 20th Design Automation Conference*, pp. 717–718.
9. Glover, C. T. and Mercer, M. R. 1988. A method of delay fault test generation. In *Proceedings of the 25th ACM/IEEE Design Automation Conference*, pp. 92–95.
10. Lala, P. K. 1996. *Practical Digital Logic Design and Testing*, Prentice-Hall, Upper Saddle River, NJ.
11. Wang, L. T., Wu, C. W. and Wen, X. 2006. *VLSI Test Principles and Architectures: Design for Testability*, Morgan Kaufmann Publishers, New York, NY.
12. Pomeranz, P. K. and Parvathala, S. P. 2007. Estimating the fault coverage of functional test sequence without fault simulation. *IEEE Computer Society*.

13. Zobrist, G. W. 1993. *VLSI Fault Modeling and Testing Techniques*, Ablex Publishing Corporation, Norwood, NJ.
14. Bushnell, M. L. and Agarwal, V. D. 2000. *Essentials of Electronic Testing*, Kluwer academic Publishers, Norwell, MA.
15. McCluskey, E. J. 1985. Built-in self-test techniques. *IEEE Design and Test of Computers* 2, no. 2: 21–28.
16. McCluskey, E. J. 1985. Built-in self-test structures. *IEEE Design and Test of Computers* 2, no. 2: 29–36.
17. Lee, K. and Breuer, M. 1992. Design and test rules for CMOS circuits to facilitate IDDQ testing of bridging faults. *IEEE Transactions on Computer-Aided Design* 11, no. 5: 659–670.

9 Nanomaterials and Applications

9.1 PREAMBLE OF NANOMATERIALS

For the past several years, the nanomaterials have been a prime focus of the researchers due to the demand for nanotechnology-based devices and equipment. The concept of nanotechnology was first introduced by the Nobel laureate in physics, Dr. Richard P. Feynman, for the popular title "There's plenty of room at the bottom" in 1959 [1]. Since then, nanomaterials have been explored and studied in detail in several domains, and experts from different fields, such as physics, chemistry, and biology, have come together for this revolutionary journey. In 1974, the term nanotechnology was first described by Norio Taniguchi to define the physical dimensions in a more precise way for the innovation of circuits, devices, and systems in the research field. Nanomaterials can be utilized for different applications in many areas by manipulating the material structure to harvest several benefits [2]. Professor Norio Taniguchi, at the Tokyo University of Science, described the "top-down approach" by predicting the miniaturization and performance improvement in very-large-scale integration (VLSI) circuit and systems, flexible electronic devices, mechanical devices, etc. Ten years later, K Eric Drexler introduced a new way to create a large object from the nanoparticles or atomic components that uses the bottom-up approach, as depicted in Figure 9.1, for next-generation nanotechnology [3]. Based on the dimensionality that changes the properties of nanomaterials, these can be categorized as 0D, 1D, 2D, and 3D nanomaterials. The term 0D can be defined when the flow of electrons is confined in all three dimensions that can be observed in quantum dots. The quantum wires (e.g., carbon nanotubes [CNTs]) are termed as 1D nanomaterials wherein the flow of electrons can be observed only in X direction; however, in 2D materials, such as thin films (e.g., graphene sheet), the movement of free electrons can be observed in the X–Y direction. Moreover, in 3D materials, the flow of free electrons can be observed in the X, Y, and Z directions [3, 4].

9.2 INTRODUCTION TO CARBON NANOTUBES

The carbon atom is primarily the key point that provides a diverse variety of components when its allotropes are used. The bonding of carbon to carbon atom in different combinations can form the sp^2 and sp^3 hybridization and produce several new compositions such as fullerenes, graphite, diamond, nanotubes, nanoribbons, etc. Of several allotropes, the CNTs are widely used based on their applicability in different areas. The use of CNTs is only possible due to their extraordinary properties and adaptability to be used remarkably in many academic and scientific areas [5]. Moreover, the applicability of carbon-based material in the nano-regime opens a new gateway to

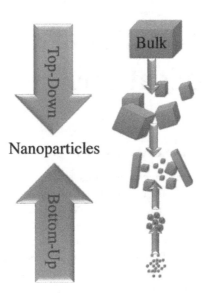

FIGURE 9.1 The process of top-down and bottom-up approach for nanoparticles.

use them in several fields such as biomedical, sensors, defense, water treatment, etc. In this chapter, the primary focus is on the carbon allotrope-based nanotubes that are widely used for interconnect application because of the limited functionality of Copper (Cu)-based interconnect at lower physical dimensions. The CNTs are an alternative nanomaterial that are used in the semiconductor industry since the ingenious discovery by Sumio Iijima in 1991 [6, 7]. The CNTs are formed by rolling up a one-atom thick planar graphene sheet into a cylindrical structure. The carbon atoms are arranged and tightly bound in honeycomb (hexagonal) lattice structure wherein each carbon atom is connected to three neighboring atoms by forming sp^2 hybridization. The cylindrical nanotubes are hollow and long having inner and outer diameter; these can be used as semiconductors, metals, and insulators depending on their defects and crystal orientation. In 1993, the single-walled carbon nanotubes (SWCNTs) were reported by Iijima as well as Bethune in separate works in the presence of cobalt-nickel crystal [6, 8]. Later in 1996, a pure SWCNT was produced by Smalley et al. using laser vaporization technique in the presence of cobalt and nickel [9].

9.2.1 The Concept of Chirality on CNT

The nature of CNTs is primarily dependent on the way in which one-atom thick graphene sheet is rolled up to produce the nanotubes. However, the growth of CNTs in a realistic scenario does not incorporate the rolled-up phenomenon of the graphene sheet. Instead, it can be produced by using several chemical vapor deposition (CVD) processes that are primarily responsible for the direct growth of CNTs vertically over the substrate [10, 11]. The physical properties of CNTs can be distinguished by varying the diameter *dia*, angle θ (in the range of 0–30°), and a pair of chiral indices (*n*, *m*). The different scenario of CNTs that exhibit armchair, zigzag, and chiral structure can be

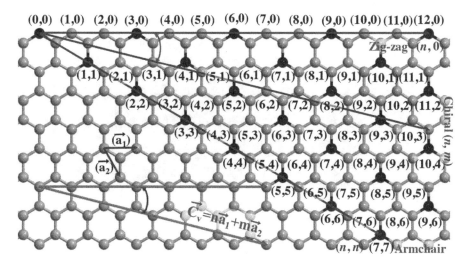

FIGURE 9.2 The illustration of honeycomb lattice structure along with electronic patterns and unit cells.

obtained based on the analysis demonstrated in Figure 9.2. The pair of chiral indices (n, m) and diameter of CNTs are primarily responsible for the nature of the structure that can be semiconductor or metallic. The combination of two lattice vectors $\vec{a_1}$ and $\vec{a_2}$ defines the chiral vector $\vec{C_v}$ as described by $\vec{C_v} = n\vec{a_1} + m\vec{a_2}$. The chiral indices (n, m) for armchair structure exhibit metallic behavior only when n and m values are equal; whereas for zigzag structure, one of the indices value must be zero (either $n = 0$ or $m = 0$) and it must satisfy $n - m = 3i$ (where i is a positive integer) condition to behave as a metal or a semiconductor. On the other hand, the CNTs are termed as chiral for different values of n and m (i.e., $n \neq m$ and $n \neq 0$). From the fabrication point of view, the growth of CNTs is a mixture of SWCNT and multi-walled CNT (MWCNT) that contains one-third metallic and two-third of semiconducting CNTs [5, 12–14].

Depending on the number of concentric cylindrical tubes, CNTs can be categorized as SWCNT and MWCNT, as demonstrated in Figure 9.3(a) and (b), respectively. If the nanotubes contain a single cylindrical wall with diameters ranging from 0.7 to 4.33 nm, they are termed as SWCNTs; however, more than one cylindrical wall with the separation of interlayer van der waals distance of 0.34 nm can be treated as MWCNTs [15]. The chiral index (n, m) is the major factor in determining the diameter of a CNT that can be defined as:

$$dia_{CNT} = \frac{\sqrt{3}a_{cc}}{\pi}\sqrt{n^2 + nm + m^2} \tag{9.1}$$

9.2.2 ELECTRONIC BAND STRUCTURE

To understand the physics behind the electronic band structure of CNTs, it is required to understand the physical structure of CNTs and graphene nanoribbons (GNRs) as discussed in Subsection 9.2.1. Depending on the chirality, the CNTs exhibit different

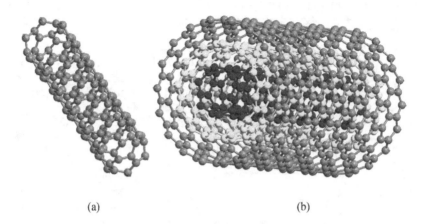

(a)   (b)

FIGURE 9.3   The illustration of (a) SWCNT and (b) MWCNT.

electronic properties such as metallic and semiconductor. The metallic behavior of CNTs is always observed for the armchair structure, whereas the zigzag structure can exhibit either semiconductor or metallic properties. Moreover, the electronic property of CNTs can also be altered by doping or creating defects on the structure to enhance the functionality based on the application for the electronic industry. The unit cell and formation of σ and π bond are depicted in Figure 9.4(a) and (b), respectively [15]. The characterization of the graphene sheet can be defined using σ and π bonds that are associated with the sp^2 hybridization. The primary contribution toward the electronic transport properties of graphene and nanotube is due to the formation of delocalized π bonding and π^* antibonding states due to the overlapping of P_z atomic orbitals [16]. However, the formation of σ bonding and σ^* antibonding occurs due to splitting of sp^2 hybrid atoms that lie far away from the Fermi energy and does not contribute to the electronic transport properties [17].

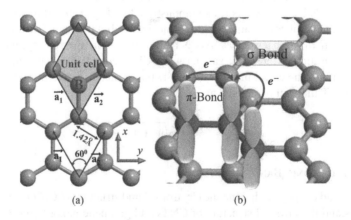

(a)   (b)

FIGURE 9.4   (a) Unit cell and (b) formation of σ- and π-bonds.

9.2.3 Brillouin Zone

In solid-state physics, the first Brillouin zone plays a major role and can be defined as Wigner–Seitz primitive cell of the reciprocal lattice that provides a brief concept about the electronic band structure. The lattice structure of the graphene sheet does not demonstrate a Bravais lattice due to different orientations of atom A and B, but it can be viewed as an underlying square (oblique) Bravais lattice using two-atom basis. In Figure 9.4(a), the $\vec{a_1}$ and $\vec{a_2}$ are the two primitive lattice vectors that define the underlying square Bravais lattice and can be represented in the Cartesian X- and Y-axis as:

$$\vec{a_1} = \left(\frac{\sqrt{3}}{2}a, \frac{1}{2}a \right), \text{ and } \vec{a_2} = \left(\frac{\sqrt{3}}{2}a, -\frac{1}{2}a \right) \tag{9.2}$$

The magnitude of a primitive vector can be defined as a, where a is the lattice constant and the lattice vector $\vec{a_1}$ and $\vec{a_2}$ can be separated with an angle of 60°. The length of the basis vector and the nearest carbon-to-carbon bonding distance (a_{cc}) can be denoted as $|\vec{a_1}| = |\vec{a_2}| = a = \sqrt{3}a_{cc}$, and $a_{cc} = 1.42086$ Å, respectively [18]. The reciprocal lattice of primitive unit cells having A and B atoms, as shown in Figure 9.4(a), can be represented using the basis vector $\vec{b_1}$ and $\vec{b_2}$ as depicted in Figure 9.5. The reciprocal lattice vector can be obtained as $\vec{b_1} = \frac{2\pi}{a} \left(\frac{1}{\sqrt{3}}, 1 \right)$ and $\vec{b_2} = \frac{2\pi}{a} \left(\frac{1}{\sqrt{3}}, -1 \right)$ by satisfying the condition $\vec{a_i}.\vec{a_j} = 2\pi\delta_{ij}$. Moreover, the higher symmetry point Γ, K, and M—as mentioned with the dotted line inside the hexagonal structure of graphene, as shown in Figure 9.5—can be obtained using a tight binding energy model. Hence,

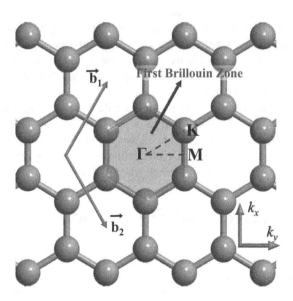

FIGURE 9.5 Reciprocal lattice of the primitive lattice vector $\vec{a_1}$ and $\vec{a_2}$ having atom A and B.

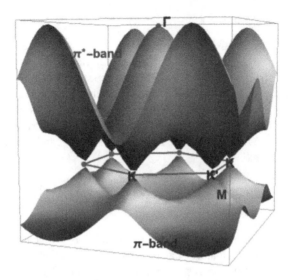

FIGURE 9.6   The energy dispersion relations of graphene [18, 19].

the electronic structure of the graphene sheet E_{GNR}^{2D} can be approximated based on the expression given below:

$$E_{GNR}^{2D}\left(k_x, k_y\right) = \pm t \left\{ 1 + 4 cos\left(\frac{\sqrt{3}}{2} ak_x\right) cos\left(\frac{1}{2} ak_y\right) + 4 cos^2\left(\frac{1}{2} ak_y\right) \right\}^{0.5} \quad (9.3)$$

where, the transfer integral is denoted as t.

Moreover, the 2D graphene sheet can be analyzed using expression (9.3) based on the E–K relationship, wherein the six corner points are highly symmetrical at the Dirac point as depicted in Figure 9.6. The energy dispersion plot consists of a conduction band that can be observed on the upper half (antibonding π^*) and valance band on the lower half (bonding π).

The point at which the conduction band and the valance band touch each other at the corners K and K' of the first Brillouin zone is known as the Fermi point and is denoted as k_F and k'_F. Moreover, the density of state (DOS) at the Fermi level becomes zero due to that graphene representing the zero bandgap at the K-point because of the higher symmetry of carbon atoms A and B.

9.3   OVERVIEW OF GRAPHENE NANORIBBON

The strip of graphene having a width less than 50 nm can be defined as GNR that can be armchair and zigzag in structure depending on the arrangement of carbon atom on the edges as depicted in Figure 9.7(a) and (b), respectively. The theoretical model of GNR was efficiently demonstrated by Mitsutaka Fujita research group in 1996. The analysis was carried out to demonstrate the impact of edge and scaling effect on GNR [20, 21]. Later in 2004, Konstantin Novoselov and Andre Geim succeeded to obtain

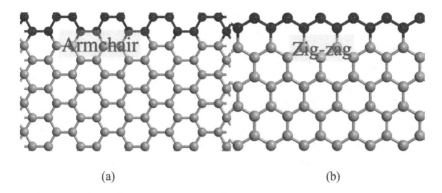

FIGURE 9.7 Two-dimensional view of (a) armchair and (b) zig-zag GNR.

one-atom-thick monolayer of carbon atoms for the first time; the same research group in 2007 reported that the number of graphene layers is primarily responsible for electronic properties of GNR [22, 23]. At the nanoscale regime, understanding the circuit behavior and electronic properties is essential to produce nanoelectronic devices based on the applications. Several research groups have reported the impact of edge roughness on GNR and change in the electronic properties theoretically and experimentally to utilize in silicon technology [24]. Recently, GNR as a novel material has gained the attention of researchers working toward the production of nanoelectronic devices based on its semiconductor properties by opening the bandgap. Moreover, the zero bandgap also motivates the researchers to replace the conventional Cu material, due to its limited functionality at the nanoscale, with metallic GNR as an interconnect application.

In addition, the GNRs can be classified as single-layered GNR (SLGNR) and multi-layered GNR (MLGNR) based on the number of GNR sheets as depicted in Figures 9.8(a) and (b), respectively [25]. A single layer of GNR sheet is termed as SLGNR, whereas more than one layer of GNR sheet sandwiched with an interlayer distance of 0.34 nm is termed as MLGNR. Moreover, in Subsection 9.2.1, it is demonstrated that the armchair CNTs are always metallic in nature, while the chiral indices (n, m) in zigzag CNTs are responsible for metallic or semiconductor

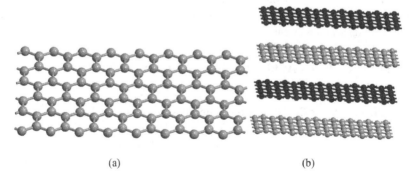

FIGURE 9.8 The structure of graphene as (a) SLGNR and (b) MLGNR.

behavior. However, the statement is conflicted in the case of GNR, as the zigzag GNRs are always metallic while armchair GNRs can behave as metallic if $N = 3i - 1$ or semiconducting if $N = 3i$ or $3i + 1$ (i is an integer) depending on the number (N) of atoms crosses the width [24, 25].

9.4   PROPERTIES OF CNT AND GNR

CNTs and GNRs can be obtained from the carbon allotropes having a unique structure that provides an extraordinary property due to the arrangement of carbon atom in sp^2 hybridization. It has been reported by several research groups that the CNTs and GNRs exhibit extreme strength and elasticity due to the tightly packed σ bonding between carbon atoms. The strength of CNT/GNR (up to 48,000 K Nm kg^{-1}) is observed as much higher than carbon steel (154 K Nm kg^{-1}) and the elastic modulus is approximately 1 TPa or 1000 Gpa in case of SWCNT in comparison to the aluminum and steel where it is just 350 GPa and 210 TPa, respectively [26, 27]. Due to extremely thin monolayer thickness and high strength and tensile property, it is possible to easily bend it. Apart from this, a good thermal behavior of CNTs and GNRs provides higher transportation of electrons along the tubes, which is also known as ballistic conduction. Experimentally, it is observed that the graphene provides improved thermal conductively of approximately 3500 W m^{-1} K^{-1} along the axis as compared to copper at room temperature. The stronger in-plane σ bonding is primarily responsible for the higher thermal conductivity below 20 K temperature and this behavioral change is used to develop nanoelectronic devices for molecular electronics, sensors and actuating devices, flexible electronics, etc. The CNTs and GNRs also exhibit a unique electrical property based on the position of atoms and chiral indices [28].

Based on the structure, CNTs can exhibit semiconducting or metallic properties. Since the miracle material graphene have metallic behavior, the nanotubes must satisfy the condition of $n - m = 3i$ (where i is an integer), and hence the armchair structure always provides metallic properties. On the other hand, based on the chirality, CNTs can be either semiconductor or metallic. Depending on the theoretical approach, the metallic nanotubes exhibit an electric current density of approximately 4×10^9 A cm^{-2}, which is 1000 times higher as compared to Al and Cu materials [29]. The electronic property primarily depends on the quality and defect-free structure of graphene. Table 9.1 displays the comparative analysis of different material properties.

GNR, graphene nanoribbon; MWCNT, multi-walled carbon nanotube; SWCNT, single-walled carbon nanotube.

9.5   FABRICATION APPROACHES FOR GRAPHENE NANOSTRUCTURE

This section describes the fabrication techniques to realize the depth insight into the nanostructure in real time. As the fabrication of electronic devices is required based on the demand of nanodevices and components such as transistors, photodetectors, and sensors in the VLSI industry, this subsection provides a general procedure of graphene fabrication techniques [30]. In the recent past, several techniques have already been developed for the fabrication of the graphene nanostructure; these can

TABLE 9.1

Important Material Properties and Their Comparison

Properties	Tungsten	Cu	SWCNT	MWCNT	GNR
Mean free path at 300 K	33	40	$>10^3$	2.5×10^4	1×10^3
Maximum current density J_{max} (A/cm²)	10^8	10^7	$>10^9$	$>10^9$	$>10^8$
Tensile strength (GPa)	1.51	0.22	22.2 ± 2.2	11–63	2.0–2.33
Temperature coefficient of resistance ($\times 10^{-3}$/K)	4.5	4	<1.1	−1.37	−1.47
Melting point (K)	3695	1357	–	2.5×10^4	1×10^3
Density (g/cm³)	19.25	8.94	1–1.4	1.7–2.1	2.09–2.33
Thermal conductivity ($\times 10^3$ W/m K)	0.173	0.385	1.75–5.8	3	3–5

be categorized as "top-down" and "bottom-up" methodologies. There are several techniques available to grow graphene such as epitaxial growth of graphene on silicon carbide, CVD, mechanical cleavage to obtain graphene from graphite, etc. The detailed fabrication process is discussed in Chapter 2.

In order to have the deep insight of the fabrication and synthesis process of graphene, the readers can refer to the content given in references [30, 31]. However, for basic understanding of the concept of obtaining the graphene, this section provides brief information on how to obtain the graphene from graphite using mechanical cleavage technique. This methodology was used for the first time by Andre Geim and Konstantin Novoselov who used the scotch tape as depicted in Figure 9.9 [32]. This approach not only provided a new gateway for the next-generation technology but also developed a new trend to enhance the performance at nanoscale. In order

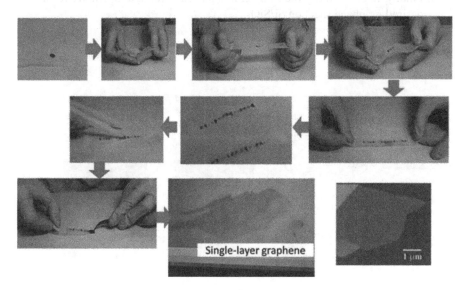

FIGURE 9.9 Single-layer graphene from graphite using mechanical cleavage [32].

to obtain a single-layered graphene, an adhesive tape is used repeatedly to peel off few layers from graphite powder. Then, adhesive tape is gently pressed on the top of the silicon wafer and is slowly peeled off. During this process, a small amount of graphite flake is deposited on the surface of the wafer [33].

Moreover, detailed information regarding the transfer of graphene on to the silicon wafer is discussed in subsequent subsections.

9.5.1   THE TRANSFER PROCESS OF GRAPHENE ON THE SI/SIO₂ SUBSTRATE

Following steps are required to elaborate the techniques to transfer the graphene onto any arbitrary substrate as depicted in Figure 9.10:

1. First, we take graphene sheet on to Cu foils and the sample will float (graphene side up) inside the etching solution of 0.1 M $(NH_4)_2S_2O_8$ [34].
2. A few drops of hexane are added into the etching solution using a syringe with the exposed face of the graphene in contact with hexane and the Cu foil exposed to the etchant solution.
3. After keeping it for approximately 12 h of etching time, the graphene sheet is trapped at the interface; surface tension for the hexane–water interface is maintained at ca. 45 mN m^{-1} to prevent the water layer from pulling the sheet apart [34].
4. Further, the fresh graphene sheet is scooped out by using Si/SiO₂ wafers.
5. To eliminate any possible contamination, the graphene sheet is again scooped out and transferred to a new hexane–pure water interface for 5 h [34].
6. Finally, in order to dry the sample, it is placed at room temperature for the desired graphene sheet on Si/SiO₂.

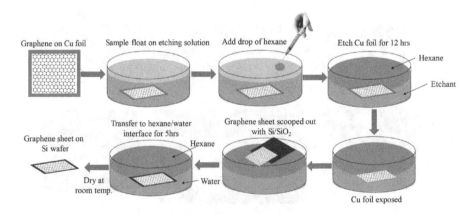

FIGURE 9.10   Fabrication process flow to transfer the graphene on Si/SiO₂ substrate [34].

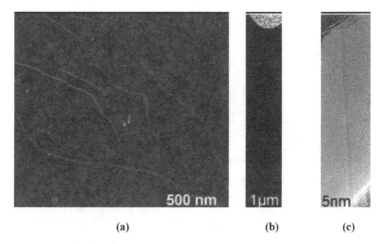

(a)   (b)   (c)

FIGURE 9.11 PECVD method used to grow SWCNT. (a) AFM view of nanotubes from low-density ferritin catalyst; (b) nanotubes grown using a catalyst particle; (c) TEM view of 1.2-nm diameter SWCNT [35].

9.5.2 CNT FABRICATIONS

For several decades, traditional approaches, such as arc discharge, CVD, and laser ablation, have been used to grow the CNTs. Out of these techniques, CVD is the most widely used methodology to grow the CNTs due to its low production cost, ease of handling, high throughput, and higher controllability over the production. The diameter, orientation, and placement of CNTs can be controlled by controlling the parameters associated with the methodology such as carbon source, reactor temperature, gas pressure, etc. [35]. Few sampled atomic force microscopy and transmission electron microscopy views of SWCNT and MWCNT have been demonstrated in Figures 9.11 and 9.12 based on the different fabrication and synthesis processes as given in references [35, 36].

9.6 APPLICATION OF NANOMATERIALS

In this section, we have described the importance of different nanomaterials based on their unique applications. The current demand for nanodevices in various fields in the VLSI domain provides a clear vision to utilize the nanomaterials to produce different electronic gadgets for different applications. Such applications include VLSI/nanointerconnects, nanosensors, nano-biosensors, etc. Following subsections describe these applications in detail.

9.6.1 GRAPHENE NANORIBBON INTERCONNECT

In the past decade, the overall performance and robustness of the interconnect has degraded with shrinking technology. It has resulted in significant attention toward the interconnect delay at the higher operating frequency. The feature size of the integrated circuit (IC) is reduced to accommodate more number of transistors and this number increases by two times per year according to Moore's law [37]. For an interconnect application, the most widely used material is Cu. With shrinking

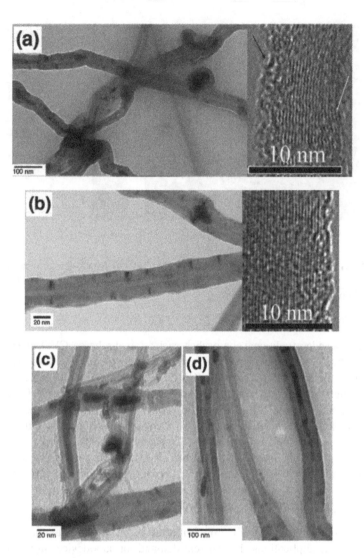

FIGURE 9.12   TEM view of MWCNT grown on a carbon paper (a) without an aluminum buffer layer, (b) with an aluminum buffer layer, (c) MWCNT grown on Si/SiO$_2$ wafer, and (d) MWCNT grown on Si/SiO$_2$–Al substrate [36].

technology and higher operating frequency, Cu exhibits several problems such as grain boundary due to the defects in crystalline structure and sidewall scattering [38]. At an advanced technology node, the electromigration (EM)-induced hillock and void formation problems have developed due to the rapid increase in the resistivity [39–43]. To overcome these issues, the research communities are motivated to adopt an alternate interconnect emerging materials such as CNT and GNR. Graphene is a thin layer of pure carbon with two-dimensional (2D) structure wherein the carbon atoms are tightly packed and bound together in a honeycomb lattice structure [44].

However, MLGNR is preferred over SLGNR as a potential interconnect material due to its higher mean free path (MFP) and more number of conducting channels.

The researchers in reference [45] examined the unique electrical and thermal behavior of graphene in terms of higher current density (~2 × 10⁹ A/cm²) and hence 2–3% of reduction in resistance has also been observed in [46]. Afterward, the improved performance of the fabricated GNR at radio frequency interconnect application was also reported in reference [47]. In addition, the authors in reference [48] demonstrated a large carrier MFP in high-quality GNR interconnects. Recently, the impact of the reduced metal–nanoribbon contact resistance is observed using hybrid interconnect compared to Cu at advanced technology [49]. Based on the above state of the art [44–49], the researchers still need to identify the solution for the challenges associated with GNR interconnect such as reduced variability, surface impurities, and contact resistance. Therefore, the analytical and simulation-based approach used until now is preferred for on-chip MLGNR interconnect.

Previously, Cui et al. [50] analyzed the transient response of the proposed model by considering capacitive and inductive coupling. The authors derived fourth-order approximation-based transfer function to investigate signal transmission behavior while neglecting the impact of interlayer tunneling conductance. Afterward, Zhao et al. [51] demonstrated an equivalent single conductor (ESC) model of tri-SLGNR interconnect lines by comparing the performance in terms of crosstalk with MLGNR. The same researchers in reference [52] extended their work and demonstrated a comparative analysis of distributed parameters and transmission characteristics of side contact MLGNR interconnects. In the proposed model, the authors neglected the impact of mutual inductance and capacitance while modeling the equivalent interconnect. Later, Qian et al. [53] proposed an ESC model of MLGNR and derived an analytical coupled transfer function at 14 nm technology to investigate the Nyquist stability criteria and crosstalk-induced delay. However, the authors neglected the interlayer transconductance and analyzed the performance based on their proposed L-models at advanced technology. Recently, Kumbhare et al. [54] demonstrated a comparative investigation of MLGNR and Cu in terms of peak noise and crosstalk-induced delay for diverse technologies. The model considered was independent of the Fermi energy and edge roughness of MLGNR. Moreover, the performance analysis between Cu and MLGNR lead the readers to the next-generation interconnect material while a detailed insight of the emerging MLGNR for future technology was missing. Considering the abovementioned facts depicted in [54], the authors in the recent research provide an extension of the previous work to deliver an in-depth analysis of next-generation interconnect materials such as MLGNR. In addition, the number of conducting channels of the proposed equivalent electrical model of MLGNR considers the edge roughness (specular constant = 0.8) and Fermi energy ($E_F = 0.6$ eV) [52] based on the technology parameters.

9.6.1.1 Geometry of MLGNR Interconnect

This section delineates the basic structure of a coupled MLGNR with the separation S between the lines, placed with a distance d (110.4, 76.8, and 52.5 nm) above the surface plane as depicted in Figure 9.13 [54]. The different dielectric materials ε_r ($\varepsilon_r = 2.25, 2.05,$ and 1.75) based on technology are used to separate the MLGNR,

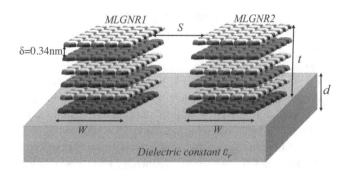

FIGURE 9.13   Geometrical structure of MLGNR interconnect as per the ITRS 2015.

wherein the line width and thickness are signified by W ($W = 32$, 22, and 14 nm) and t ($t = 32$, 22, and 14 nm) [50], respectively.

For an accurate analysis, the interconnect line is considered a square segment, and the width of MLGNR is mapping with its equal thickness [50]. All the layers of MLGNR are stacked on each other with an interlayer separation of $\delta = 0.34$ nm due to the van der Waals force [55, 56]. The thickness t and interlayer spacing δ can be used to obtain the overall number of layers as:

$$N_{layer} = 1 + Int \left\lfloor \frac{t}{\delta} \right\rfloor \tag{9.4}$$

The number of conducting channel N_{ch} of an individual layer in MLGNR can be obtained by the sum of all the conduction n_c and valance n_v subbands associated with all the electrons and holes, respectively [57, 58]. It can be calculated as:

$$N_{ch} = \sum_{j=1}^{n_c} \left[1 + e^{(E_i - E_F)/k_B T} \right]^{-1} + \sum_{j=1}^{n_v} \left[1 + e^{(E_i + E_F)/k_B T} \right]^{-1} \tag{9.5}$$

where j represents the positive integer (i.e., $j = 1, 2, 3, \ldots$). k_B, E_F, and T represent the Boltzmann's constant, Fermi energy, and room temperature, respectively.

Using equation (9.5), the overall conducting channels of a rough-edged MLGNR at different Fermi energy are presented in Figure 9.14. It is inferred that the number of conducting channels drastically increases for higher values of Fermi energy at room temperature.

The number of conducting channels $N_{ch} = 23$, 15, and 10 are analytically obtained at 32, 22, and 14 nm technology, respectively for a fixed $E_F = 0.6$ eV [52, 59]. However, irrespective of technology nodes, the MFP of rough-edged MLGNR can be considered as $\lambda_{eff} = 419$ nm at room temperature for $E_F = 0.6$ eV [52]. Based on the interconnect structure and the conducting channels, the equivalent multi-conductor transmission line (MTL) and ESC models of rough-edged MLGNR are delineated in the following subsections.

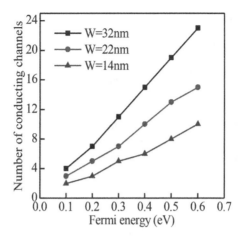

FIGURE 9.14   Total number of conducting channels corresponds to width and Fermi energy for different technology nodes.

9.6.1.2   Equivalent MTL Model of MLGNR Interconnect

This subsection demonstrates an accurate MTL model using a *Pi*-type distributed network, as shown in Figure 9.15. The quantitative values of parasitics in per unit length (*p.u.l.*) are determined at different technology nodes based on International Technology Roadmap for Semiconductor (ITRS) 2015 [60] using the interconnect geometry, as shown in Figure 9.13. Depending on the fabrication house, the interconnect line at the interface exhibits imperfect contact R_{mc} and quantum resistance R_q, respectively. Therefore, the lumped resistance R_{fix} is considered accurately in the proposed model as $R_{fix} = R_{mc} + R_q = 3.2$ kΩ at both the terminal interface.

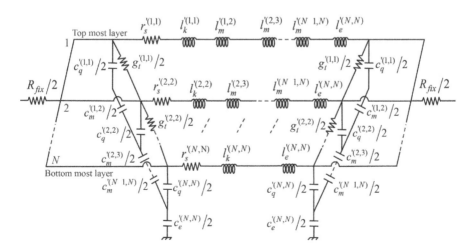

FIGURE 9.15   An *RLC*-based MTL model of rough-edged MLGNR lines [54].

For longer interconnect, an individual layer of rough-edged MLGNR unveils scattering resistance (r_s') due to defects, edged roughness, and impurities [61–63] that can be represented in *p.u.l.* as:

$$r_s' = \frac{h/2e^2}{N_{ch}\lambda_{eff}} \tag{9.6}$$

where h and *e* denote the Plank's constant and the electron charge, respectively. The effective MFP λ_{eff} depends on the defects λ_d and edged roughness λ_n associated with scattering effect for each layer of MLGNR and can be calculated according to Matthiessen's rules as:

$$\frac{1}{\lambda_{eff}} = \frac{1}{\lambda_d} + \frac{1}{\lambda_n} \tag{9.7}$$

where the associated parameters of effective MFP can be calculated based on the expression given in reference [52]. Further, along with the scattering resistance, an individual layer of MLGNR exhibits a distributed kinetic (l_k') and magnetic inductance (l_e'). It occurs due to the kinetic energy stored in the individual layer and the magnetic field associated with MLGNR. On the other hand, the distributed capacitance also comprises a quantum (c_q') and electrostatic (c_e') capacitance, respectively [54, 64]. These can be calculated as:

$$l_k' = \frac{h/4e^2}{N_{ch}v_F} \tag{9.8}$$

$$c_q' = 4e^2 N_{ch}/h v_F \tag{9.9}$$

$$l_e' = \frac{\mu_0 d}{W} \tag{9.10}$$

$$c_e' = \frac{\varepsilon_0 \varepsilon_r W}{d} \tag{9.11}$$

where the Fermi velocity of graphene is considered as $v_F \approx 8 \times 10^5$ m/s [50]. The μ_0 is free space magnetic permeability and ε_0 is the electrostatic permittivity. Here, the impact of mutual inductance (l_m') and capacitance (c_m') along with the tunneling conductance (g_t') are considered between adjacent layers of interconnect lines to improve the accuracy of the proposed equivalent model and can be calculated as:

$$l_m' = \frac{\mu_0 \delta}{W} \tag{9.12}$$

$$c_m' = \frac{\varepsilon_0 W}{\delta} \tag{9.13}$$

$$g_t' = \sigma \pi W \tag{9.14}$$

where the graphene sheet exhibits the normalized tunneling conductivity denoted as σ.

9.6.1.3   ESC Model of MLGNR Interconnect

The layers of a rough-edged MLGNR in the MTL model are assumed to be parallel as shown in Figure 9.15. Further, it can be simplified to an equivalent electrical model to reduce the computational effort, as shown in Figure 9.16. The driver-interconnect-load (DIL) setup consists of a driver capacitance (C_{tr}) and resistance (R_{tr}) along with a lumped resistance (R_{fix}) equally divided on either side of the interconnect wire of two contacts. For peak noise and crosstalk-induced delay demonstration, the driver parameters are accurately considered based on the ITRS 2015 benchmark [60] at different technology nodes. The coupled MLGNR line of the ESC model terminated with capacitance load C_L [43, 60].

Figure 9.16 primarily comprises of equivalent scattering resistance $(r'_{s,esc})$ that is calculated by dividing the total number of layers in equation (9.6) as:

$$r'_{s,esc} = \frac{r'_s}{N_{layer}} \tag{9.15}$$

In addition, the overall effective inductance (l'_{esc}) can be calculated using the sum of the magnetic inductance $(l'_{e,esc})$ and the kinetic inductance $(l'_{k,esc})$ of the ESC model. The $l'_{k,esc}$ can be computed by dividing the number of layers into equation (9.8) as:

$$l'_{esc} = l'_{k,esc} + l'_{e,esc} \tag{9.16}$$

$$l'_{k,esc} = \frac{l'_k}{N_{layer}} \tag{9.17}$$

$$l'_{e,esc} = l'_e \tag{9.18}$$

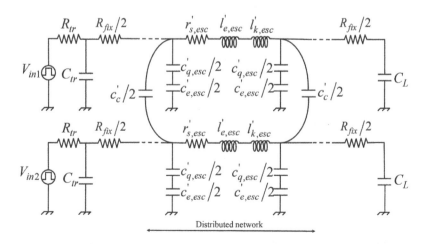

FIGURE 9.16   An *RLC*-based simplified ESC model of rough-edged MLGNR lines [54].

The impact of mutual inductance l'_m (in the range of pH/μm) can be ignored between the adjacent layers of MLGNR expressed in equation (9.12) that is insignificant as compared to the kinetic inductance l'_k (usually in the range of nH/μm) [43].

The MLGNR interconnect also exhibits total equivalent capacitance that comprises three capacitive components—quantum, electrostatics, and mutual coupling capacitance. The equivalent quantum capacitance $c'_{q,esc}$ can be observed due to the density of electronic state that is experimentally validated in [65]. The equivalent electrostatic capacitance $c'_{e,esc}$ can be observed owing to the potential difference between the lowermost layer and the surface plane. The *p.u.l.* c'_{esc} capacitive components of the MLGNR can be calculated as:

$$c'_{q,esc} = c'_q N_{layer} \tag{9.19}$$

$$c'_{e,esc} = c'_e \tag{9.20}$$

$$c'_{m,esc} = \left(\frac{1}{c'^{(1,2)}_m} + \frac{1}{c'^{(2,3)}_m} + \dots + \frac{1}{c'^{(N-1,N)}_m} \right) \tag{9.21}$$

where $c'_{m,esc}$, c'_m, and N denote the equivalent mutual coupling capacitance, the mutual coupling capacitance, and the number of layers of MLGNR interconnect.

9.6.1.4 Validation of MTL and ESC Model

In order to demonstrate the accuracy, the proposed MTL and equivalent ESC models of MLGNR in Figures 9.15 and 9.16 are validated using a coupled interconnect line. The performance is validated in terms of crosstalk-induced delay.

Figure 9.17 represents the delay under induced crosstalk at advanced technology for line spacing of 5 and 20 nm, respectively. It is evident that at 5-nm spacing,

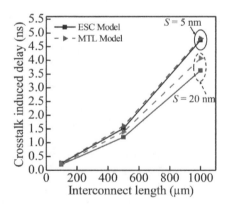

FIGURE 9.17 At 14-nm node, the MTL and ESC models are validated for different spacing between coupled rough-edged MLGNR in terms of delay.

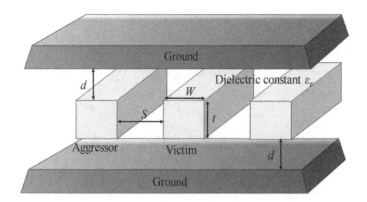

FIGURE 9.18 ITRS-based interconnect geometry represents a cross-sectional view.

the crosstalk-induced delay using MTL is in good agreement with the ESC one. However, for a more spacing of 20 nm, an approximate difference of 0.91% in performance can be observed using the proposed ESC and the MTL models of a rough-edged MLGNR. Therefore, the proposed ESC model is accurate for analyzing the crosstalk-induced delay at shrinking technology.

9.6.1.5 Simulation Setup of MLGNR

This subsection presents an ITRS 2015 [60] based interconnect structure in the form of a cross-sectional view as depicted in Figure 9.18. Depending on the interconnect structure (Figure 9.18), the peak noise and crosstalk-induced delay are demonstrated by considering a coupled DIL as depicted in Figure 9.19. Out of these two interconnects, one acts as the aggressor while another as a victim [39]. Using ITRS 2015 benchmark [60], accurate physical parameters (i.e., W, t, S, ε_r and d) for a given interconnect technology are considered. Using these parameters, the quantitative values of interconnect parasitics are described in Table 9.2 to investigate the delay and peak noise under the influence of crosstalk. The performance of coupled rough-edged MLGNR is primarily governed by the coupling capacitance (c_c') that is the function

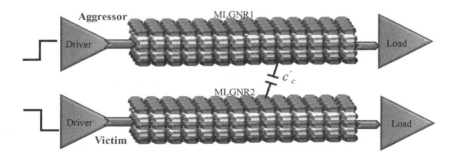

FIGURE 9.19 Simulation setup of rough-edged MLGNR for out-phase transition using coupled DIL.

TABLE 9.2

Interconnect Parasitics at Different Technology Nodes

Parasitics of MLGNR Interconnect	Quantitative Value at Technology Node (nm) of		
	32	22	14
Driving resistance R_{tr} (kΩ)	12.13	16.67	18.33
Driving capacitance C_{tr} (fF)	0.081	0.049	0.030
Scattering resistance $r'_{s,esc}$ (Ω/μm)	14.13	31.67	73.53
Magnetic inductance $l'_{e,esc}$ (pH/μm)	1.96	2.85	4.48
Kinetic inductance $l'_{k,esc}$ (pH/μm)	3.70	8.29	19.25
Quantum capacitance $c'_{q,esc}$ (fF/μm)	422.09	188.34	81.13
Electrostatic capacitance $c'_{e,esc}$ (aF/μm)	12.46	8.56	5.45

of width and space between interconnects and is summarized in Table 9.3 based on the *p.u.l.* parameters as [54, 66]:

$$c'_c = \left(\frac{\varepsilon}{4}\right) X \left[\sqrt{1-(1+2W/S)^{-2}} \right] \tag{9.22}$$

$$X(p) = \begin{cases} \dfrac{6.28}{\ln \dfrac{2\left[\left(1+\left(1-p^2\right)^{0.25}\right)\right]}{1-\left(1+p^2\right)^{0.25}}}, & 0 \le p \le 0.7071 \\[4ex] 0.637 \ln \dfrac{2\left(1+\sqrt{p}\right)}{1-\sqrt{p}}, & 0.7071 \le p \le 1 \end{cases} \tag{9.23}$$

9.6.1.6 Crosstalk-Induced Delay Analysis

The performance in terms of crosstalk can be determined based on the switching characteristics of coupled interconnect lines. Based on different switching scenario,

TABLE 9.3

Coupling Capacitance (c'_c) at Different Technology Nodes

Line Spacing (nm)	Quantitative Value of c'_c between the Coupled MLGNR in aF/μm at Technology Node (nm) of		
	32	22	14
5	24.86	22.73	18.51
10	20.97	18.99	16.76
15	18.83	16.98	14.92
20	17.40	15.64	13.73

the crosstalk can be characterized as the dynamic and functional type. With reference to a ground line, when an aggressor line is excited with a pulse input, the victim line experiences a voltage spike as a peak noise known as the functional crosstalk. Besides, dynamic crosstalk can be experienced when the victim and the aggressor line are excited by simultaneous pulse inputs. Under this category, the input switching for aggressor and victim wires is either in the same or in the opposite phase [67–69].

When opposite switching transition (out-phase) is applied to coupled interconnect lines, the impact of crosstalk is larger as compared to the in-phase switching. This phenomenon can be observed due to Miller's effect that is primarily governed by the coupling capacitance. Consequently, the delay and peak noise under the influence of crosstalk are comparatively investigated in the following subsections for different interconnect spacing and advanced technology.

9.6.1.6.1 Impact of Spacing between Coupled Interconnect

This subsection demonstrates the performance of coupled rough-edged MLGNR interconnect for different line spacing and interconnect length. The analysis is carried out for the proposed model using the DIL setup as depicted in Figure 9.19 at 32, 22, and 14 nm technology. Using interconnect parasitics introduced in Tables 9.2 and 9.3, an efficient analysis is observed in terms of crosstalk-induced delay for diverse space as depicted in Figures 9.20(a)–(c), at 32, 22, and 14 nm technology, respectively. Furthermore, it is evident that the signal transition period (delay under crosstalk) increases for interconnect lengths ranging from 100 to 1000 µm while it is reduced for more space between the coupled interconnects. It is evident due to the rise in the quantitative value of interconnect parasitics for longer interconnects that significantly increases the delay under the influence of crosstalk. However, at a particular technology node, the impact of crosstalk reduces for more interconnect spacing due to a reduced coupling capacitance as observed from equation (9.22). As demonstrated in Table 9.3, the quantitative value of coupling capacitance reduces for more spacing using 14 nm technology in comparison to 32 nm and 22 nm, that is the major cause for improved crosstalk-induced delay. Apart from this, an accurate analysis of the coupled interconnect is required, as crosstalk is an important parameter metrics for on-chip digital VLSI that causes unintentional triggering, glitches, and noise at the output of digital ICs.

At the nanoscale dimension, the peak noise drastically affects the performance that causes triggering of the aggressor line with input pulse while the victim line is grounded. It is observed from Table 9.4 that the unintentional peak voltage significantly increases for lower technology and longer interconnect, respectively, due to p.u.l. increase in the parasitic values of rough-edged MLGNR. The reason behind an increased resistive and inductive parasitics (Table 9.2) is the reduced number of MLGNR layers and conducting channels at advanced technology. However, it can also be observed that for the lower value of quantum and electrostatic capacitance, the overall equivalent capacitance $c'_{esc} = \left(1/c'_{q,esc} + 1/c'_{e,esc}\right)^{-1}$ increases. Therefore, the peak noise increases at the advanced technology due to the collective impact of resistive and capacitive values.

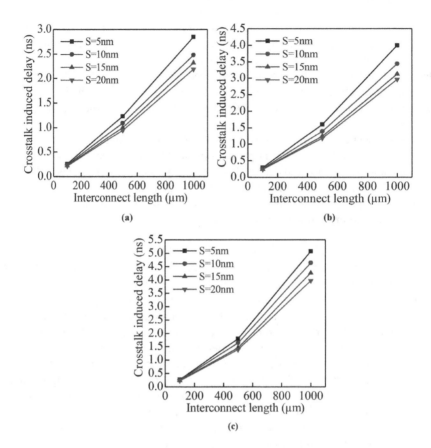

FIGURE 9.20 Crosstalk delay of rough-edged MLGNR at (a) 32 nm, (b) 22 nm, and (c) 14 nm technology, respectively, for different length and interconnect spacing.

9.6.1.6.2   Impact of Shrinking Node on Crosstalk Effect

Recent demand for faster switching and higher operating frequency motivates researchers to investigate the impact of the shrinking node on the crosstalk effect.

TABLE 9.4

Peak Voltage with Respect to Variation in Space at Different Technology Nodes

Lengths (μm)	Peak Voltage (mV) at S = 5 nm			Peak Voltage (mV) at S = 10 nm			Peak Voltage (mV) at S = 15 nm			Peak Voltage (mV) at S = 20 nm		
	32 nm	22 nm	14 nm	32 nm	22 nm	14 nm	32 nm	22 nm	14 nm	32 nm	22 nm	14 nm
100	205	241	254	186	220	247	174	207	232	165	197	221
500	277	325	369	256	303	359	243	288	345	234	279	336
1000	297	341	387	276	321	377	262	308	363	252	299	352

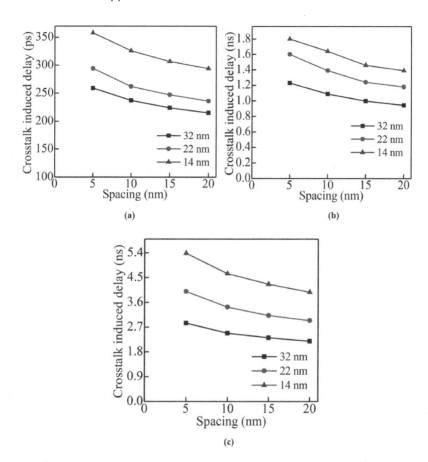

FIGURE 9.21 Crosstalk delay of rough-edged MLGNR for (a) 100 μm (b) 500 μm, and (c) 1000 μm length, respectively, for different interconnect spacing and technology.

In this subsection, the effect of scaled-down technology on the crosstalk-induced delay is demonstrated in Figure 9.21(a)–(c) for diverse interconnect lengths. As shown in Table 9.3, it can be perceived that an advanced technology increases the interconnect parasitics in *p.u.l.* due to a reduced number of MLGNR layers that significantly affects the performance, and hence the delay becomes prominent. Consequently, the effect of the technology node can be observed in Figure 9.21(a)–(c), wherein the delay increases as technology is scaled down and reduces for more spacing under the influence of crosstalk.

In addition, Table 9.5 summarizes the percentage reduction in crosstalk-induced delay at scaled-down technology for rough-edged MLGNR interconnect. It is evident that the percentage improvement in delay under crosstalk is more for 14 nm node as compared to 22 nm with respect to 32 nm technology at the longer interconnect. The main cause behind this effect is the reduced interconnect width at lower technology that drastically reduces the coupling capacitance as can be observed from Table 9.5.

TABLE 9.5

Percentage Reduction of Crosstalk Delay at Advanced Technology nodes

	% Reduction of Crosstalk Delay at 22 nm and 14 nm *w.r.t.* 32 nm Technology for Diverse Interconnect Spacing (nm) of							
	S = 5		S = 10		S = 15		S = 20	
Lengths (µm)	22 nm	14 nm	22 nm	14 nm	22 nm	14 nm	22 nm	14 nm
100	11.90	27.65	9.54	27.30	9.31	27.03	8.89	26.87
500	23.12	34.22	21.58	33.53	19.51	31.64	18.62	31.59
1000	28.57	47.02	27.69	46.55	25.64	45.53	25.51	44.83

Therefore, the overall delay under crosstalk is enriched by 45.1% at 14 nm whereas the improvement is only 26.7% for 22 nm compared to the 32 nm technology at 1000 µm wire length. Hence, a rough-edged MLGNR can be considered an evolving material for next-generation interconnect application at the shrinking node.

9.6.2 CARBON NANOTUBE-BASED INTERCONNECT

At present, the technology requirement primarily focuses on the device functionality, scalability, and increased interconnect speed. In order to meet the requirements, an MWCNT bundle and MLGNRs can be considered promising interconnect materials due to their outstanding electrical, mechanical, and chemical properties [69]. The growth of an edged MLGNR is preferred over an MWCNT due to its planar structure. Moreover, the fabrication process of rough-edged MLGNR is favored as a smooth edge with a narrow width is harder to obtain due to its uncontrollable cutting direction that requires high-resolution lithography [70]. Moreover, as per the fabrication house [71], a densely packed MWCNT bundle is hard to achieve due to: (1) the nonuniform growth of CNT bundles, (2) distributed CNT diameters, and (3) difficulty in handling the position of an individual CNT at a desired location inside the bundle. Therefore, the researchers [70–72] currently focus on the fabrication of edged MLGNR compared to a densely packed MWCNT bundle.

Previously, Sarto et al. [73] analyzed the current distribution along an edged MLGNR and compared it with MWCNT without considering an identical interconnect structure and imperfect contact resistance. Later, Majumder et al. [56] accurately accounted for the metal–nanotube contact resistance to analyze the propagation delay of an MLGNR and single MWCNT while considering their different shapes. Furthermore, the same research group in reference [67] demonstrated the impact of propagation delay and crosstalk for the random and spatial arrangement of mixed CNT bundle (MCB) considering their accurate rectangular shapes. However, the analysis was restricted to only MCB and hence MLGNR was not taken into account. Afterward, Kumar et al. [74] extended the analysis of delay and relative stability for MLGNR and MWCNT interconnects by arbitrarily considering only a few number of shells and layers, respectively. However, an accurate geometrical structure was neglected for equivalent delay and area demonstration.

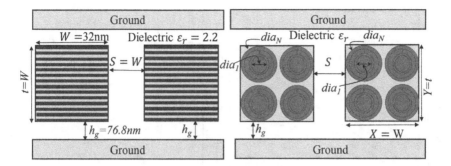

FIGURE 9.22 Cross-sectional view of MLGNR/bundled MWCNT interconnect based on ITRS.

9.6.2.1 Interconnect Model

This subsection presents an MTL model of MLGNR and bundled MWCNT interconnects based on their structure shown in Figure 9.22. The number of layers, N_{layer}/shells, N_{shell} in MLGNR/MWCNT in Figure 9.22 primarily depends on the width W, diameter dia, thickness t, and an interlayer (or intershell) distance δ and can be expressed as:

$$N_{layer(or\ shell)} = 1 + Integer[t/\delta] \qquad (9.24)$$

where $t = [(dia_N - dia_1)/2]$ for MWCNT with outermost and innermost diameters of dia_N and dia_1, respectively.

As per the ITRS requirement [60], the research work considers an equal width and thickness for rough-edged MLGNR and MWCNT bundle. Depending on the geometry, the equivalent Pi-type MTL model of MLGNR/bundled MWCNT is shown in Figure 9.15, which is further simplified as an ESC (Figure 9.23) to reduce the computational effort [75]. The ESC model is primarily driven by a resistive driver with resistance R_{tr} and capacitance C_{tr} and terminated by a load capacitance, C_L. The interconnect parasitics shown in Figures 9.15 and 9.23 are obtained based

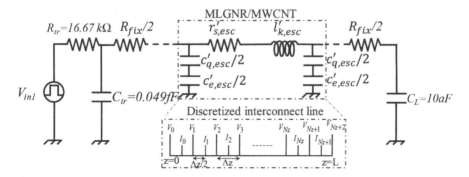

FIGURE 9.23 Discretized interconnect of the equivalent electrical model of MLGNR and MWCNT interconnects.

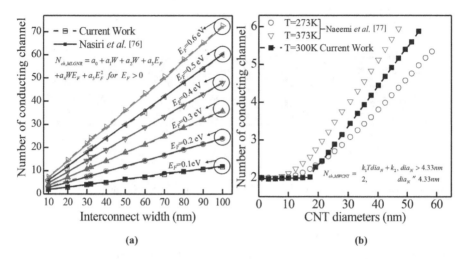

FIGURE 9.24 Validation of conducting channel associated with (a) MLGNR and (b) MWCNT for different interconnect width and diameter, respectively.

on the total number of conducting channels associated with each layer and shell of MLGNR and MWCNT, respectively, and can be expressed using expression (9.5). Figure 9.24(a) and (b) depicts the validation of the number of conducting channels of MLGNR and MWCNT with the experimental results given in references [76] and [77] that primarily differ by only 1.21 and 4.5%, respectively.

The near and far end of the interconnect line in Figures 9.15 and 9.23 is presented by a lumped resistance R_{fix}, which can be assumed as the sum of imperfect contact and quantum resistances of MLGNR and MWCNT bundle. The equivalent scattering resistance $r'_{s,esc}$ can be obtained by considering all the layers or shells in parallel in the MTL and can be expressed as [56]

$$r'_{s,esc} = r'_{s,MLGNR}/N_{layer} \text{ and } r'_{s,esc} = r'_{s,MWCNT}/N_{MWCNT} \tag{9.25}$$

$$r'_{s,MLGNR} = \left(h/2e^2\right)/N_{ch,MLGNR}\lambda_{eff} \text{ and } r'_{s,MWCNT} = \left(h/2e^2\right)/N_{total}\lambda_{eff} \tag{9.26}$$

where

$$N_{total} = \sum_{i=1}^{N} N_{ch,MWCNT} \tag{9.27}$$

Here h, e, N_{MWCNT}, and N_{total} denote the Planck's constant, electron charge, number of MWCNTs in a bundle, and a total number of conducting channels, respectively. The effective MFP (λ_{eff}) primarily depends on the edge roughness properties and can be considered 419 nm and 1 μm for edged MLGNR and MWCNT, respectively. The p.u.l. scattering and quantum resistance (obtained through expressions (9.25) to (9.26)) are validated with experimental results given in references [78] and [79] as

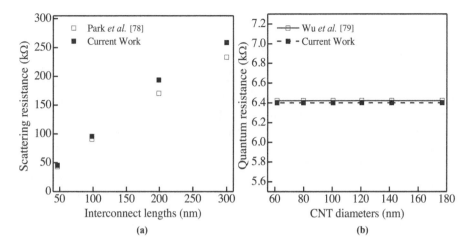

FIGURE 9.25 (a) Validation of scattering resistance with experimental data [78], (b) validation of quantum resistance with experimental data [79].

presented in Figure 9.25(a). It is observed that the scattering resistances are in good agreement with the experimental data with an average deviation of only 5.59 and 9.93% at local and global interconnects, respectively. Similarly, the experimental measurements of quantum resistance also demonstrate a close approximation with the industry-standard simulation data as observed in Figure 9.25(b).

The equivalent kinetic inductance of Figure 9.15 primarily represents the kinetic energy stored in each layer or shell in MLGNR or MWCNT and the *p.u.l.* $l'_{k,esc}$ can be expressed as [56]:

$$l'_{k,esc} = l'_{k,MLGNR}/N_{layer} \text{ and } l'_{k,esc} = l'_{k,MWCNT}/N_{total} \tag{9.28}$$

$$l'_{k,MLGNR} = (h/2e^2)/N_{ch,MLGNR}v_F \text{ and } l'_{k,MWCNT} = (h/2e^2)/N_{ch,MWCNT}v_F \tag{9.29}$$

Here, the impact of magnetic inductance (l'_e) is neglected in the ESC model of Figure 9.23 due to its negligible effect (in the order of pH/μm) in comparison with the kinetic inductance (in the order of nH/μm). The *RLC* model of Figure 9.23 also considers a quantum capacitance ($c'_{q,esc}$) that can be defined as the equivalent charge stored in each layer and shell of MLGNR and MWCNT, respectively. However, the charge stored between the bottom most layer in MLGNR and bundled MWCNT can be considered as $c'_{e,esc}$. The *p.u.l.* $c'_{q,esc}$ and $c'_{e,esc}$ can be obtained as [56]:

$$c'_{q,esc} = c'_{q,MLGNR} \times N_{layer} \text{ and } c'_{q,esc} = c'_{q,MWCNT} \times N_{total} \tag{9.30}$$

$$c'_{q,MLGNR} = 2e^2 N_{ch,MLGNR}/hv_F \text{ and } c'_{q,MWCNT} = 2e^2 N_{ch,MWCNT}/hv_F \tag{9.31}$$

$$c'_{e,esc(MLGNR)} = \frac{\varepsilon_0\varepsilon_r}{h_g} \text{ and } c'_{e,esc(MWCNT)} = \frac{2\pi\varepsilon_0\varepsilon_r}{cosh^{-1}\left[(h_g + dia_N/2)/dia_N\right]} \tag{9.32}$$

where $\varepsilon_r = 2.2$ [74], ε_0, and μ_0 are dielectric constant, electrostatic permittivity, and the magnetic permeability of free space, respectively. It is observed from model expressions (9.28) to (9.31) that the kinetic inductance and quantum capacitance of MLGNR and MWCNT depend on $N_{\text{ch,MLGNR(MWCNT)}}$, quantum resistance, and Fermi velocity (v_F). The $N_{\text{ch,MLGNR(MWCNT)}}$ and quantum resistance are already validated with experimental data as presented in Figure 9.24(a), (b) and Figure 9.25(b). The Fermi velocity $\approx 8 \times 10^5$ m/s is considered accurately from experimental data [80] in order to determine the quantitative value of kinetic inductance and quantum capacitance. Therefore, the model expressions (9.28) to (9.31) possess a good approximation with the standard experimental results.

In order to maintain the accuracy, the MTL model considers the impact of width-dependent tunneling conductance (g_t'), mutual inductance (l_m'), and coupling capacitance (c_m') between adjacent layers and shells of MLGNR and bundled MWCNT, respectively. However, these parasitics have a negligible effect in the equivalent model of Figure 9.23 as the proposed ESC and MTL models are in good agreement (less than 4.6% deviation at 1000 μm interconnect length) in terms of their delay performance. Using the proposed ESC model, the quantitative values of interconnect parasitics are summarized in Table 9.6.

9.6.2.2 Performance Comparison

This subsection analyzes the equivalent number of MWCNTs $(N_{\text{shell}} = 10)$ in a densely packed bundle for a fixed area of MLGNR at 22 nm technology node. This comparison has been performed for an equivalent delay of edged MLGNR by using a DIL setup. The interconnect line in DIL primarily represents the equivalent electrical model of MLGNR and bundled MWCNT. Industry-standard HSPICE simulations are performed to obtain the equivalent delay that is further validated against the analytical finite-difference time domain (FDTD) approach as described in the following subsections.

TABLE 9.6

Interconnect Parasitics for MLGNR and Bundled MWCNT Based on ITRS

Interconnect Parasitics	Rough-Edged MLGNR $N_{\text{layer}} = 95$	Approximate Number of MWCNTs in a Bundle as					
		$N_{\text{MWCNT}} = 7$	$N_{\text{MWCNT}} = 29$	$N_{\text{MWCNT}} = 105$	$N_{\text{MWCNT}} = 371$	$N_{\text{MWCNT}} = 1891$	$N_{\text{MWCNT}} = 9491$
N_{total}	2185	54	223	806	2848	14517	72863
R_{fix} (kΩ)	6.4	6.4	6.4	6.4	6.4	6.4	6.4
$r_{s,esc}'$ (Ω/μm)	14.13	3.38	0.28	0.02	1.71×10^{-3}	66.21×10^{-6}	2.62×10^{-6}
$l_{k,esc}'$ (pH/μm)	3.70	105.35	36.33	10.03	2.84	0.56	0.01
$c_{q,esc}'$ (fF/μm)	422.09	103.81	1247.21	16.35×10^3	204.12×10^3	5.31×10^6	133.59×10^6
$c_{e,esc}'$ (aF/μm)	12.46	39.26	39.26	39.26	39.26	39.26	39.26

9.6.2.2.1   FDTD-Based Delay Model

Figure 9.23 also presents an FDTD discretization of uniform interconnect line of MLGNR/MWCNT. Further, the interconnect line is divided into N_z section of each having a length of Δz, and the total time is divided into N_t segments of each step of Δt. The interconnect line can be described by telegrapher's equations as:

$$\frac{d}{dz}V(z,t)+r'_{s,esc}I(z,t)+l'_{k,esc}\frac{d}{dt}I(z,t)=0 \tag{9.33a}$$

$$\frac{d}{dz}I(z,t)+c'_{esc}\frac{d}{dt}V(z,t)=0 \tag{9.33b}$$

where

$$c'_{esc}=\left(1/c'_{q,esc}+1/c'_{e,esc}\right)^{-1} \tag{9.34}$$

Using discretized interconnect line of MLGNR/MWCNT into equations (9.33a) and (9.33b), the voltage and current expression can be obtained as:

$$V_k^{n+1}=[E][F]V_k^n+\{[E]/\Delta z\}\left[I_{k-1}^{n+1/2}-I_k^{n+1/2}\right] \tag{9.35}$$

$$I_k^{n+3/2}=[B][D]I_k^{n+1/2}+[B]\left[V_k^{n+1}-I_{k+1}^{n+1}\right] \tag{9.36}$$

where $[E]=(c'_{esc}/\Delta t)^{-1}$, $[F]=(c'_{esc}/\Delta t)$, $[B]=\left[(\Delta z/\Delta t)l'_{k,esc}+(\Delta z/2)r'_{s,esc}\right]^{-1}$, and $[D]=\left[(\Delta z/\Delta t)l'_{k,esc}-(\Delta z/2)r'_{s,esc}\right]$

At the near-end terminal, the voltage and current are represented by V_0 and I_0, respectively as:

$$V_0^{n+1}=\left\{1/\left[(C_{tr}/\Delta t)+1/R_{tr}\right]\right\}\left[(C_{tr}/\Delta t)V_0^n+(U/R_{tr})V_s^{n+1}-I_0^{n+1}\right] \tag{9.37}$$

$$I_0^{n+1}=\left(1/R_{fix}\right)\left[V_0^{n+1}-V_1^{n+1}\right] \tag{9.38}$$

At the far-end terminal, the voltage and current are represented as:

$$V_{N_Z+1}^{n+1}=\frac{EF}{\left[U+\frac{E}{2\Delta zR_{fix}}\right]}V_{N_Z}^1+\frac{E}{\Delta z\left[U+\frac{E}{2\Delta zR_{fix}}\right]}\left[\frac{V_{N_Z+2}^{n+1}}{2R_{fix}}+I_{N_Z}^{n+1/2}-\frac{1}{2}I_{N_Z+1}^n\right] \tag{9.39}$$

$$I_{N_Z+1}^{n+1}=\left(1/R_{fix}\right)\left[V_{N_Z+1}^{n+1}-V_{N_Z+2}^{n+1}\right];\; V_{N_Z+2}^{n+1}=V_{N_Z+2}^n+\left(\Delta t/C_L\right)I_{N_Z+1}^n \tag{9.40}$$

The propagation delay can be obtained by solving the voltage and current expression (9.33) to (9.40) in space and time for MLGNR/bundled MWCNT interconnect.

9.6.2.2.2 Simulation Model

This subsection presents a real-time simulation-based approach to compare the performance of bundled MWCNT and edged MLGNR interconnects. Using the DIL setup of an edged MLGNR at 22 nm technology node, an equivalent delay is obtained for a different number of MWCNTs (with $N_{shell} = 10$) in a densely packed bundle, as shown in Figure 9.26(a). It is observed that for a fixed number of MLGNR layers ($N_{layer} = 95$) at 22 nm technology node, the number of MWCNTs in a bundle significantly increases for longer interconnects. It is due to the impact of resistive and capacitive parasitics that primarily dominate the overall delay of an interconnect line. In the proposed model, the total resistance is dominated by the imperfect contact resistance (in the range of kΩ) of MLGNR/bundled MWCNT over the *p.u.l.* scattering resistance (in the range of Ω). Due to the consideration of a similar value

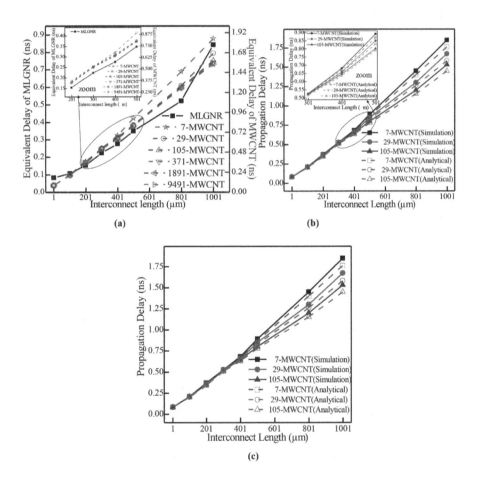

FIGURE 9.26 Propagation delay of (a) an edged MLGNR and bundled MWCNT for an equivalent delay, (b) bundled MWCNT (for $N_{MWCNT} = 7, 29, 105$), and (c) bundled MWCNT (for $N_{MWCNT} = 371, 1891, 9491$) using analytical and simulation-based models at different interconnect length, respectively.

of R_{fix} for both MLGNR and bundled MWCNT (as in Table 9.6), the overall resistance has a negligible impact on the interconnect delay. However, the equivalent capacitance (c'_{esc}) has a large impact on the overall delay of an interconnect line. As observed in Table 9.6, the c'_{esc} of an edged MLGNR/bundled MWCNT is primarily governed by the electrostatic capacitance, $c'_{e,esc}$, (in the order of aF/μm) over the equivalent quantum capacitance, $c'_{q,esc}$, (in the order of fF/μm). Due to the smaller value of $c'_{e,esc}$, the equivalent capacitance, c'_{esc} of edged MLGNR primarily dominates over that of the bundled MWCNT as observed from equation (9.34) and Table 9.6. Hence, on average, a rough-edged MLGNR illustrates 44.5% lesser delay compared to a bundled MWCNT for global interconnect lengths.

In order to demonstrate the accuracy, the real-time simulation model is compared with the analytical FDTD-based DIL of bundled MWCNT, as shown in Figure 9.26(b) and (c). Figure 9.26(b) demonstrates the accuracy of the model for $N_{MWCNT} = 7, 29$, and 105, whereas the delay of $N_{MWCNT} = 371, 1891$, and 9491 is shown in Figure 9.26(c). The propagation delay obtained through the proposed FDTD model differs by 2.75% for fewer MWCNTs ($N_{MWCNT} = 105$), whereas the difference is only 1.92% for more MWCNTs ($N_{MWCNT} = 9491$) in a densely packed bundle. The proposed analytical model also demonstrates its accuracy with the simulation results by 1.97 and 3.53% for local and global interconnects, respectively. Table 9.7 demonstrates the impact of area for edged MLGNR and bundled MWCNT on the basis of the equivalent delay performance at 22 nm technology node as observed in Figure 9.26(a). As the number of MWCNTs in a densely packed bundle is lesser compared with the number of layers in MLGNR for local interconnects, the MLGNR consumes 30.17% more area compared to a bundled MWCNT interconnect. However, the area requirement is increased by 97.98% for an MWCNT bundle in comparison to an edged MLGNR for global interconnects as observed in Table 9.7.

TABLE 9.7

Percentage Improvement in the Overall Area for an Edged MLGNR *w.r.t.* Bundled MWCNT at 22 nm Technology Node

N_{MWCNT} and Bundle Area for a Fixed $N_{layer} = 95$ and MLGNR Area ($W \times t$) = 0.001024 μm² to Obtain an Equivalent Delay at Different Interconnect Lengths of

Interconnect length (μm)	N_{MWCNT}	MWCNT Bundle area (μm²) = $X \times Y$[a]	% IMPROVEMENT in the area for MLGNR with Respect to Bundled MWCNT
1	7	0.0004	−58.00%
100	29	0.0010	−2.34%
200	105	0.0061	83.21%
500	371	0.0211	95.14%
800	1891	0.1054	99.02%
1000	9491	0.5274	99.80%

[a] where $X = \left[\sum_{i=1}^{N_x} dia_{N_i} + \sum_{i=1}^{N_x-1} \delta_i \right]$ and $Y = \left[\sum_{i=1}^{N_y} dia_{N_i} + \sum_{i=1}^{N_y-1} \delta_i \right]$

Therefore, a rough-edged MLGNR can outperform a bundled MWCNT with more number of shells for longer interconnects. Moreover, the production of MLGNR is more straightforward than an MWCNT bundle as per the requirement by recent fabrication houses [72]. Hence, a rough-edged MLGNR can be considered the most promising interconnect material than MWCNT bundle in terms of fabrication and performance at global interconnect.

9.6.3 NANOSENSOR

During the past decades, the use of sensors in daily lives of people has increased, but some areas are still unreachable for sensing technology which require advanced technological nanodevices [81]. The function of nanodevices should include the ability to sense, and based on the sensing data it should react accordingly in real time. In recent technology, the nanomaterial-based nanosensors and devices are growing rapidly that are used to measure the pulse rate, heart functions, glucose level, blood pressure, and many more health-related parameters [82]. The development of nanotechnology has unwrapped wide areas of utilizing the technology in the medical field by integrating multiple sensors in a single chip for advanced levels of observations. Carbon allotrope-based nanocomposite, nanowires, and nanotubes have become the most popular nanomaterial to improve the sensitivity of nanoelectronic devices [83]. This subsection provides a brief introduction about a few applications based on the nanomaterials.

9.6.3.1 Flexible Sensor

Nanomaterials are used to create a flexible sensor to measure the amount of deflection or bending curvature. The flexible sensor primarily depends on the variation of resistive components that are mounted on the surface [84, 85]. The amount of resistance deflection can be measured on the basis of the bending of the flexible sensor. The flexible sensor is used in several areas of research such as environment monitoring, security systems, robotics, body health monitoring systems, smartwatches, smart cell phones, display, computer interface, etc. [86]. The production of nanosensing devices needs to consider several factors such that it can be cost-effective, lightweight, transparent, soft, biocompatible, and easy to fabricate. The key applications such as flexible panels, electronic skin, wearable sensors, structural health monitoring, and space flight based on the flexible sensor are demonstrated in Figure 9.27 [87–89].

9.6.3.2 Nanosensor for Biomedical Applications

In recent technology, the nanomaterial-based nanosensors and devices are growing rapidly that are used to measure the pulse rate, heart functions, glucose level, blood pressure, and many more health-related diseases. The development of nanotechnology has opened a new way of utilizing the technology in the medical field by integrating multiple sensors in a single chip for advanced levels of observations. Carbon allotrope-based nanocomposites, nanowires, and nanotubes have become the most popular nanomaterials to improve the sensitivity of nanodevices [90]. At the nanoscale dimension, the property of the material is used to develop efficient sensors

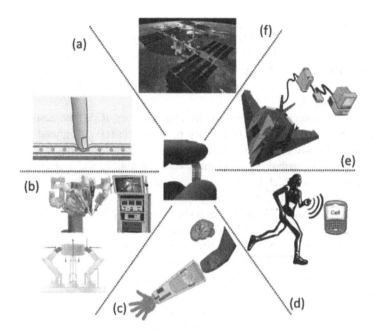

FIGURE 9.27 Various applications of the flexible sensor [88].

that utilize the change in melting point, electrical conductivity, magnetic permeability, large surface-to-volume ratio of the nanoparticles. Figure 9.28 represents a general steps of the processing of the nanosensor devices [91, 92].

9.6.4 NANOMATERIAL-BASED COMBAT JACKETS

In order to protect the extremely valuable life of the soldiers, it is required to produce a bulletproof combat jackets that are lighter and cheaper than previous ones using nanomaterials. Mostly, the body armor is made by the Kevlar material that can be classified as aramid substance. However, the recent research proved that the graphene can be used as a novel material for body armors. The graphene is extremely powerful and 10 times a better substitute for Kevlar.

A recent breakthrough for graphene applications is in materials for bulletproof vests developed by combining CNTs with GNRs. In 2012, Prof. Min Kyoon Shin, Bommy Lee, and their research group [93] from the University of Wollongong,

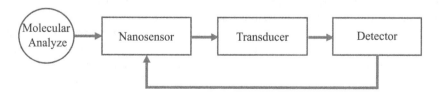

FIGURE 9.28 The process of nanosensors.

Australia, developed a new graphene-based material that is tougher than the substances such as spider silk and Kevlar that are widely employed in making of bulletproof vests. They found that the new material toughness (1000 J g^{-1}) is much better than that of the spider silk (165 J g^{-1}) and Kevlar (78 J g^{-1}). The toughness of the material can be increased by a proper combinational ratio of reduced graphene oxide flakes and CNTs [93].

In 2014, the material scientist Edwin Thomas from Rice University, Houston, and Jae-Hwang Lee from the University of Massachusetts, Amherst, worked on graphene-based body armor and performed the impact test using small graphene sheet. They fired microbullets of 9 mm toward a graphene sheet of 200 × 200 mm at a speed of 3 km/s, which is much faster than the speed fired by the AK-47 assault rifle [94]. This test concluded that the graphene is successfully able to distribute the stress of the bullets over a wide area at a speed of up to 2200 m/s.

In 2015, Prof. Shantanu Bhowmik, head of the Research and Projects at school of Engineering, Amrita Vishwa Vidyapeetham, Coimbatore, India, and his research group (MadhavDatta and Siva Kumar) worked on a bulletproof jacket using the ultramodern lightweight thermoplastic Carbon Fabric technology for soldiers [95].

In 2017, scientists Yang Gao, Tengfei Cao, Filippo Cellini, and their group from Advanced Science Research Center, City University of New York investigated that the ultrathin layer of graphene can exhibit extreme hardness and transverse stiffness property as like as a diamond. The researchers demonstrated that the transverse stiffness is about 400 GPa that is much higher than that of the bare substrate for two layers [96].

Since the miracle material graphene, which has one atom thick carbon atoms arranged in honeycomb hexagonal configuration, was discovered, this scientifically important nanomaterial-based body armor has been used in the field of military and defense for enhancing the protection of the soldiers. Additionally, it is extremely light in weight and has an extraordinary property to conduct heat and electricity with improved efficiency. Moreover, high stiffness, strong elasticity, and better tensile strength make this material superior for body armor.

9.6.5 Nano-Biosensors for Drug Delivery

In the field of medicine, the identification of diseases at the early stage is a challenging task. In order to overcome these challenges and provide cost-effective solutions, the different disciplines of science and technology are working together [97, 98]. The main purpose is to provide nanodiagnostics with more accurate data such that the diseases can be identified at early stages. Using cost-effective nano-biosensors, various diseases, such as cancers, infectious diseases, immunodeficiency, neurological disorders, etc., can be examined accurately [99]. Advancement and technological development provide a platform to establish the nanomedicine that includes molecular imaging, drug delivery, and regenerative systems. At present, DNA-based biosensors are the most fascinating and perspective sensors. A DNA-based nano-biosensor consists of a transmitter, transducer, and processing unit to receive accurate data, as shown in Figure 9.29 [100].

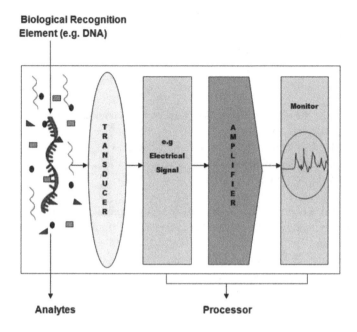

Biological Recognition
Element (e.g. DNA)

Analytes   Processor

FIGURE 9.29 A generalized description for nano-biosensor [97].

9.7 SUMMARY

This chapter can be summarized as follows:

- A brief preamble to nanomaterials was provided such that the reader can obtain enough information related to nanomaterials and their applicability in science and technology.
- Carbon allotrope-based nanomaterials, such as CNTs and GNRs, were explored along with their extraordinary properties.
- Based on the current demand in the VLSI industry, electrical modeling of emerging nanomaterials was demonstrated to observe the performance of interconnect at an advanced technology node.
- Several nanosensor-based devices were also described such as flexible sensors and biosensors for biomedical applications.
- Finally, the chapter concludes with the short, long, and MCQ-type questions to check how effectively the reader has gained information from this chapter.

9.8 MULTIPLE-CHOICE QUESTIONS

1. The prefix "nano" comes from a …
   a. French word meaning billion
   b. Greek word meaning dwarf
   c. Spanish word meaning particle
   d. Latin word meaning invisible

2. Who first used the term nanotechnology and when?
   a. Richard Feynman, 1959
   b. Norio Taniguchi, 1974
   c. Eric Drexler, 1986
   d. Sumio Iijima, 1991

3. Which of these historical works of art contain nanotechnology?
   a. Lycurgus cup
   b. Medieval stained glass windows in churches
   c. Damascus steel swords
   d. All of the above

4. Richard Feynman is often credited with predicting the potential of nanotechnology. What was the title of his famous speech given on December 29, 1959?
   a. There is a tiny room at the bottom
   b. Things get nanoscopic at the bottom
   c. Bottom? What bottom?
   d. There is plenty of room at the bottom

5. What is graphene?
   a. A new material made from carbon nanotubes
   b. A one-atom thick sheet of carbon
   c. Thin film made from fullerenes
   d. A software tool to measure and graphically represent nanoparticles

6. What is the general name for the class of structures made of rolled-up carbon lattices?
   a. Nanorods
   b. Nanotubes
   c. Nanosheets
   d. Fullerrods

7. Which of the following is an example of top-down approach for the preparation of nanomaterials?
   a. Gas phase agglomeration
   b. Molecular self-assembly
   c. Mechanical grinding
   d. Molecular beam epitaxy

8. Which of the following is an example of bottom-up approach for the preparation of nanomaterials?
   a. Etching
   b. Dip pen nano-lithography
   c. Lithography
   d. Erosion

9. The carrier transport in graphene goes through π-bands that result from...
   a. *sp²* hybridization of valence electrons
   b. *sp³* hybridization of valence electrons
   c. Unhybridized *s*-orbital valence electrons
   d. Unhybridized *p*-orbital valence electrons

10. What does the "chirality" (n, m) denote for carbon nanotubes (CNT)?
   a. If the CNT is single-walled or multi-walled
   b. If the CNT is insulating or metallic
   c. A direction that the graphene sheet is rolled up to form a tube
   d. A direction that the CNT extends along

11. Carbon nanotubes are the sheets of graphite about the thickness
   a. 0.1 nm
   b. 0.2 nm
   c. 0.3 nm
   d. 0.4 nm

12. Carbon nanotubes were first observed in the year
   a. 1992
   b. 1991
   c. 1990
   d. 1993

13. Graphene is a
   a. Zero-dimensional (0D) material
   b. One-dimensional (1D) material
   c. Two-dimensional (2D) material
   d. Three-dimensional (3D) material

14. Graphene
   a. Is more than 100 times stronger than steel
   b. Is about as stiff as diamond
   c. Has the highest electron mobility of all electronic materials
   d. All of the above

15. What is the smallest atom that can pass through a sheet of defect-free, single-layer graphene?
   a. Cerium
   b. Oxygen
   c. Most atom can pass
   d. Not even the smallest atoms (helium, hydrogen) can pass

9.9 SHORT ANSWER QUESTIONS

1. List out three challenges associated with carbon nanotube-based interconnect.
2. Define nanotechnology.
3. List out the applications of nanotechnology in electronics.
4. Define fringing capacitance and area capacitance.
5. Define nanostructured materials.
6. List any four day-to-day live commercial applications of nanotechnology.
7. Define Brillouin zone.
8. What is crosstalk?
9. How to reduce crosstalk on coupled interconnect?
10. What are the factors that influence the performance of interconnect?

9.10 LONG ANSWER QUESTIONS

1. Explain in detail electrical, thermal, and mechanical properties of CNTs and GNRs.
2. Explain the bottom-up approach of the synthesis of nanomaterials.
3. Explain the importance of a flexible sensors.
4. How to obtain the carbon nanotubes from the graphene sheet? Explain in detail.
5. What are the challenges faced by the researchers in nanotechnology?
6. Write a short note on graphene nanoribbon.
7. Write a short note on the band structure of graphene.
8. Explain Miller's effect in detail.
9. Explain any one nanosensor used in biomedical application.
10. Why nanomaterial-based combat jacket is important for defense and military?

REFERENCES

1. Feynman, R. P. 1960. There's plenty of room at the bottom. *Engineering and Science* 23, no. 5: 22–36.
2. Taniguchi, N. 1974. On the basic concept of "Nano-Technology". *Proceedings of the International Conference on Production Engineering* Tokyo, Part II, Japan Society of Precision Engineering, Tokyo, pp. 5–10.
3. Drexler, K. E. 2006. Engines of Creation 2.0 — The Coming Era of Nanotechnology, *Wowio*.
4. Siegel, R. W. 1993. Synthesis and properties of nanophase materials. *Materials Science and Engineering: A* 168, no. 2: 189–197.
5. Das, S. 2013. A review on carbon nano-tubes—A new era of nanotechnology. *International Journal of Emerging Technology and Advanced Engineering* 3, no. 3: 774–781.
6. Iijima, S. and Ichihashi, T. 1993. Single-shell carbon nanotubes of 1-nm diameter. *Nature* 363, no. 6430: 603–605.
7. Iijima, S. 1991. Helical microtubes of graphitic carbon. *Nature* 354, no. 6348: 56–58.
8. Bethune, D., Kiang, C. and de Vries, M., *et al.* 1993. Cobalt-catalysed growth of carbon nanotubes with single-atomic-layer walls. *Nature* 363, 605–607.

9. Hafner, J. H., Bronikowski, M. J., Azamian, B. R., Nikolaev, P., Rinzler, A. G., Colbert, D. T., Smith, K. A. and Smalley, R. E. 1998. Catalytic growth of single-wall carbon nanotubes from metal particles. *Chemical Physics Letters* 296, no. 1–2: 195–202.

10. Kataura, H., Kumazawa, Y., Maniwa, Y., Ohtsuka, Y., Sen, R., Suzuki, S. and Achiba, Y. 2000. Diameter control of single-walled carbon nanotubes. *Carbon* 38, no. 11–12: 1691–1697.

11. Bandow, S., Asaka, S., Rao, Y. A. M., Grigorian, L., Richter, E. and Eklund, P. C. 1998. Effect of the growth temperature on the diameter distribution and chirality of single-wall carbon nanotubes. *Physical Review Letters* 80: 3779–3782.

12. Sinnott, S. B. and Andreys, R. 2001. Carbon nanotubes: Synthesis, properties, and applications. *Critical Reviews in Solid State and Materials Sciences* 26, no. 3: 145–249.

13. Guan, L., Suenaga, K. and Iijima, S. 2008. Smallest carbon nanotube assigned with atomic resolution accuracy. *Nano Letters* 8, no. 2: 459–462.

14. Wang, X., Li, Q., Xie, J., Zhong, J., Wang, J., Li, Y., Jiang, K. and Fan, S. 2009. Fabrication of ultralong and electrically uniform single-walled carbon nanotubes on clean substrates. *Nano Letters* 9, no. 9: 3137–3141.

15. Javey, A. and Kong, J. 2009. *Carbon Nanotube Electronics*, Springer, Berlin.

16. Wallace, P. R. 1947. The band theory of graphite. *Physical Review Letters* 71, no. 9: 622–634.

17. Minto, E. 2004. Tuning the band structure of carbon nanotubes. Ph.D. Thesis, Cornell University.

18. Saito, R., Dresselhaus, M. S. and Dresselhaus, G., *et al.* 2004. *Physical Properties of Carbon Nanotubes*, Imperial College Press, London, UK.

19. Fathi, D. 2011. A review of electronic band structure of graphene and carbon nanotubes using tight binding. *Journal of Nanotechnology*, Article 471241: 1–6.

20. Fujita, M., Wakabayashi, K., Nakada, K. and Kusakabe, K. 1996. Peculiar localized state at zigzag graphite edge. *Journal of the Physical Society of Japan* 65, no. 7: 1920–1923.

21. Nakada, K., Fujita, M., Dresselhaus, G. and Dresselhaus, M. S. 1996. Edge state in graphene ribbons: Nanometer size effect and edge shape dependence. *Physical Review Letters* 54, no. 24: 17954–17961.

22. Wakabayashi, K., Fujita, M., Ajiki, H. and Sigrist, M. 1999. Electronic and magnetic properties of nanographite ribbons. *Physical Review B* 59, no. 12: 8271–8282.

23. Novoselov, K. S., Geim, A. K., Morozov, S. V., Jiang, D., Zhang, Y., Dubonos, S. V., Grigorieva, I. V. and Firsov, A. A. 2004. Electric field effect in atomically thin carbon films. *Science* 306, no. 5696: 666–669.

24. Liying, J., Li, Z., Xinran, W., Georgi, D. and Hongjie, D. 2009. Narrow graphene nanoribbons from carbon nanotubes. *Nature* 458, no. 7240: 877–880.

25. Yang, X., Dou, X., Rouhanipour, A., Zhi, L., Räder, H. J. and Müllen, K. 2008. Two-dimensional graphene nanoribbons. *Journal of the American Chemical Society* 130, no. 13: 4216–4217.

26. Sinnott, S. B. and Andrews, R. 2001. Carbon nanotubes: Synthesis, properties and applications. *Critical Reviews in Solid State and Materials Sciences* 26, no. 3: 145–249.

27. Popov, V. N. 2004. Carbon nanotubes: Properties and application. *Materials Science and Engineering R: Reports* 43: 61–102.

28. Burke, P. J. 2002. Lüttinger liquid theory as a model of the gigahertz electrical properties of carbon nanotubes. *IEEE Transactions on Nanotechnology* 1, no. 3: 129–144.

29. Pop, E., Mann, D. A., Goodson, K. E. and Dai, H. 2007. Electrical and thermal transport in metallic single-wall carbon nanotubes on insulating substrates. *Journal of Applied Physics* 101, no. 9: 093710–1093710-10.

30. Nazarpour, S. and Waite, S. R. 2016. *Graphene Technology from Laboratory to Fabrication*, Wiley-VCH, New York, NY.

31. Novoselov, K. S., Geim, A. K., Morozov, S. V., Jiang, D., Zhang, Y., Bubonos, S. V., Grigorieva, I. V. and Firsov, A. A. 2004. Electric field effect in atomically thin carbon films. *Science* 306, no. 5696: 666–669.

32. Yi, M. and Shen, Z. 2015. A review on mechanical exfoliation for the scalable production of graphene. *Journal of Materials Chemistry A* 3, no. 22: 11700–11715.

33. Blake, P., Hill, E. W., Castro Neto, A. H., Novoselov, K. S., Jiang, D., Yang, R., Booth, T. J. and Geim, A. K. 2007. Making graphene visible. *Applied Physics Letters* 91, no. 6: 063124.

34. Zhang, G., Güell, A., Kirkman, P., Lazenby, R., Miller, T. and Unwin, P. 2016. Versatile polymer-free graphene transfer method and applications. *ACS Applied Materials & Interfaces* 8, no. 12: 8008–8016.

35. Li, Y., Mann, D. and Rolandi, M., *et al.* 2004. Preferential growth of semiconducting single-walled carbon nanotubes by a plasma enhanced CVD method. *Nano Letters* 4, no. 2: 317–321.

36. Liu, H., Zhang, Y., Arato, D., Li, R., M´erel, P. and Sun, X. 2008. Aligned multi-walled carbon nanotubes on different substrates by floating catalyst chemical vapor deposition: Critical effects of buffer layer. *Surface and Coatings Technology* 202, no. 17: 4114–4120.

37. Moore, G. E. 1998. Cramming more components onto integrated circuits. *Proceedings of IEEE* 86, no. 1: 82–85.

38. Kaushik, B. K., Majumder, M. K. and Kumar, V. R. 2014. Carbon nanotube based 3-D interconnects—A reality or a distant dream. *IEEE Circuits Systems and Magazine* 14, no. 4: 16–35.

39. Majumder, M. K., Kumar, J. and Kaushik, B. K. 2015. Process-induced delay variation in SWCNT, MWCNT, and mixed CNT interconnects. *IETE Journal of Research* 61, no. 5: 533–540.

40. Davis, J. A. and Meindl, J. D. 2003. *Interconnect Technology and Design for Gigascale Integration*, Springer, London.

41. Kreupl, F., Graham, A. P., Duesberg, G. S., Steinhögl, W., Liebau, M., Unger, E. and Hönlein, W. 2002. Carbon nanotubes in interconnect applications. *Microelectronic Engineering* 64, no. 1–4: 399–408.

42. Srivastava, N. and Banerjee, K. 2005. Performance analysis of carbon nanotube interconnects for VLSI applications. *Proceedings of IEEE/ACM International Conference on Computer-Aided Design*, San Jose, CA, USA, pp. 383–390.

43. Li, H., Yin, W. Y., Banerjee, K. and Mao, J. F. 2008. Circuit modeling and performance analysis of multi-walled carbon nanotube interconnects. *IEEE Transactions on Electron Devices* 55, no. 6: 1328–1337.

44. Chen, X., Akinwande, D. and Lee, K. J., *et al.* 2010. Fully integrated graphene and carbon nanotube interconnects for gigahertz high-speed CMOS electronics. *IEEE Transactions on Electron Devices* 57, no. 11: 3137–3143.

45. Behnam, A., Lyons, A. S. and Bae, M. H., *et al.* 2012. Transport in nanoribbon interconnects obtained from graphene grown by chemical vapor deposition. *Nano Letters* 12, no. 9: 4424–4430.

46. Kang, C. G., Lim, S. K. and Lee, S., *et al.* Effects of multi-layer graphene capping on Cu interconnects. *Nanotechnology* 24, no. 11: 115707-1–115707-5.

47. Nguyen, P. D., Nguyen, T. C., Huynh, A. T. and Skafidas, S. 2014. High frequency characterization of graphene nanoribbon interconnects. *Materials Research Express* 1, no. 3: 035009-1–035009-10.

48. Politou, M., Asselberghs, I. and Soree, B., *et al.* 2016. Single- and multilayer graphene wires as alternative interconnects. *Microelectronic Engineering* 156, no. C: 131–135.

49. Lee, C. S., Shin, K. W. and Song, H. J., *et al.* 2018. Fabrication of metal/graphene hybrid interconnects by direct graphene growth and their integration properties. *Advanced Electronic Materials* 4, no. 6: 1700624-1–1700624-8.

50. Cui, J. P., Zhao, W. S. and Yin, W. Y. 2012. Signal transmission analysis of multilayer graphene nano-ribbon (MLGNR) interconnects. *IEEE Transactions on Electromagnetic Compatibility* 54, no. 1: 126–132.

51. Zhao, W. S. and Yin, W. Y. 2012. Signal integrity analysis of graphene nano-ribbon (GNR) interconnects. *IEEE Electrical Design of Advanced Packaging and Systems Symposium (EDAPS)*, Taipei, pp. 227–230.

52. Zhao, W. S. and Yin, Y. Y. 2014. Comparative study on multilayer graphene nanoribbon (MLGNR) interconnects. *IEEE Transactions on Electromagnetic Compatibility* 56, no. 3: 638–645.

53. Qian, L., Xia, Y. and Shi, G. 2016. Study of crosstalk effect on the propagation characteristics of coupled MLGNR interconnects. *IEEE Transactions on Nanotechnology* 15, no. 5: 810–819.

54. Kumbhare, V. R., Paltani, P. P. and Majumder, M. K. 2018. Future of graphene based interconnect technology—A reality or a distant dream. *Proceedings of 5th IEEE Uttar Pradesh Section International Conference on Electrical, Electronics and Computer Engineer (UPCON)*, Gorakhpur, pp. 1–7.

55. Xu, C., Li, H. and Banerjee, K. 2009. Modeling, analysis, and design of graphene nanoribbon interconnects. *IEEE Transactions on Electron Devices* 56, no. 8: 1567–1578.

56. Majumder, M. K., Kukkam, N. R. and Kaushik, B. K. 2014. Frequency response and bandwidth analysis of multi-layer graphene nanoribbon and multi-walled carbon nanotube interconnects. *IET Micro & Nano Letters* 9, no. 9: 557–560.

57. Fathi, D., Forouzandeh, B., Mohajerzadeh, S. and Sarvari, R. 2009. Accurate analysis of carbon nanotube interconnects using transmission line model. *IET Micro & Nano Letters* 4, no. 2: 116–121.

58. Nasiri, S. H., Moravvej-Farshi, M. K. M. and Faez, R. 2010. Stability analysis in graphene nanoribbon interconnects. *IEEE Electron Device Letters* 31, no. 12: 1458–1460.

59. Jiang, J., Kang, J., Cao, W. and Banerjee, K. 2015. UCSB graphene nanoribbon interconnect compact model.

60. International Technology Roadmap for Semiconductors (ITRS). [Online], http://www.itrs.net.

61. Areshkin, D. A., Gunlycke, D. and White, C. T. 2007. Ballistic transport in graphene nanostrips in the presence of disorder: Importance of edge effects. *Nano Letters* 7, no. 1: 204–210.

62. Hwang, E. H., Adam, S. and Sarma, S. D. 2007. Carrier transport in two-dimensional graphene layers. *Physical Review Letters* 98, no. 18: 186806-1–186806-4.

63. Yan, J., Zhang, Y., Kim, P. and Pinczuk, A. 2007. Electric field effect tuning of electron-phonon coupling in graphene. *Physical Review Letters* 98, no. 16: 166802-1–166802-4.

64. Nishad, A. K. and Sharma, R. 2014. Analytical time-domain models for performance optimization of multilayer GNR interconnects. *IEEE Journal of Selected Topics in Quantum Electronics* 20, no. 1: 17–24.

65. Xia, J. L., Chen, F., Li, J. H. and Tao, N. J. 2009. Measurement of the quantum capacitance of graphene. *Nature Nanotechnology* 5: 505–509.

66. Stellari, F. and Lacaita, A. L. 2000. New formulas of interconnect capacitances based on results of conformal mapping method. *IEEE Transactions on Electron Devices* 47, no. 1: 222–231.

67. Majumder, M. K., Kaushik, B. K. and Manhas, S. K. 2014. Analysis of delay and dynamic crosstalk in bundled carbon nanotube interconnects. *IEEE Transactions on Electromagnetic Compatibility* 56, no. 6: 1666–1673.

68. Majumder, M. K., Das, P. K. and Kaushik, B. K. 2014. Delay and crosstalk reliability issues in mixed MWCNT bundle interconnects. *Microelectronic Reliability* 54, no. 11: 2570–2577.
69. Hazra, A. and Basu, S. 2018. Graphene nanoribbon as potential on-chip interconnect material—A review. *Carbon* 4, no. 49: 1–27.
70. Wang, X., Ouyang, Y., Jiao, L., Wang, H., Xie, L., Wu, J., Guo, J. and Dai, H. 2011. Graphene nanoribbons with smooth edges behave as quantum wires. *Nature Nanotechnology* 6, no.9: 563–567.
71. Kim, S., Kulkarni, D. D., Rykaczewski, K., Henry, M., Tsukruk, V. V. and Fedorov, A. G. 2012. Fabrication of an ultralow-resistance ohmic contact to MWCNT–Metal interconnect using graphitic carbon by electron beam-induced deposition (EBID). *IEEE Transactions on Nanotechnology* 11, no. 6: 1223–1230.
72. Chappanda, K. N., Batra, N. M., Holguin-Lerma, J. A., Costa, P. M. F. J. and Younis, M. I. 2017. Fabrication and characterization of MWCNT-based bridge devices. *IEEE Transactions on Nanotechnology* 16, no. 6: 1037–1046.
73. Sarto, M. S. and Tamburrano, A. 2010. Comparative analysis of TL models for multilayer graphene nanoribbon and multiwall carbon nanotube interconnects, *IEEE International Symposium of Electromagnetic Compatibility*, Fort Lauderdale, FL, pp. 212–217.
74. Kumar, V. R., Majumder, M. K., Alam, A., Reddy, N. and Kaushik, B. K. 2015. Stability and delay analysis of multi-layered GNR and multi-walled CNT interconnects. *Journal of Computational Electronics, Springer* 14: 611–618.
75. Rabaey, J. M., Chandrakasan, A. and Nikolic, B. 2017. *Digital Integrated Circuits: A Design Perspective*, 2nd ed., Prentice-Hall, Englewood Cliffs, NJ, USA.
76. Nasiri, S. H., Faez, R. and Moravvej-Farshi, M. K. 2012. Compact formulae for number of conduction channels in various types of graphene nanoribbons at various temperatures. *Modern Physics Letters B* 26, no. 01: 115004–115005.
77. Naeemi, A. and Meindl, J. D. 2006. Compact physical models for multiwall carbonnanotube interconnects. *IEEE Electron Device Letters* 27, no. 5: 338–340.
78. Park, J. Y., Rosenblatt, S., Yaish, Y., Sazonova, V., Ustunel, H., Braig, S., Arias, T. A., Brouwer, P. W. and Mceuen, P. L. 2003. Electron-phonon scattering in metallic singlewalled carbon nanotubes. *Nano Letters* 4, no. 3: 517–520.
79. Wu, W., Krishnan, S., Yamada, T., Sun, X., Wilhite, P., Wu, R., Li, K. and Yang, C. 2009. Contact resistance in carbon nanostructure via interconnects. *Applied Physics Letters* 94, no. 16: 163113-1–163113-3.
80. Hwang, C., Siegel, D. A., Ma, S. K., Regan, W., Ismach, A., Zhang, Y., Zettl, A. and Lanzara, A. 2012. Fermi velocity engineering in graphene by substrate modification. *Scientific Reports* 2, no. 590: 1–4.
81. Dahman, Y. 2017. Nanotechnology and functional materials for engineers. *Nanosensors* 67–91.
82. Kim, S. J., Choi, S. J., Jang, J. S., Cho, H. J. and Kim, I. D. 2017. Innovative nanosensor for disease diagnosis. *Accounts of Chemical Research* 50, no. 7: 1587–1596.
83. Rong, G., Tuttle, E. E., Neal Reilly, A. and Clark, H. A. 2019. Recent developments in nanosensors for imaging applications in biological systems. *Annual Review of Analytical Chemistry* 12, no. 1: 109–128.
84. Pantelopoulos, A. and Bourbakis, N. G. 2010. A survey on wearable sensor-based systems for health monitoring and prognosis. *IEEE Transactions on Systems, Man, Cybernetics: System* 40, no. 1: 1–12.
85. Bandokar, A. J., Jeang, W. J., Ghaffari, R. and Rogers, J. A. 2019. Wearable sensors for biochemical sweat analysis. *Annual Review of Analytical Chemistry* 12: 1–22.

86. An, B., Ma, Y., Li, W., Su, M., Li, F. and Song, Y. 2016. Three-dimensional multi-recognition flexible wearable sensor via graphene aerogel printing. *Chemical Communications* 52, 10948–10951.

87. Apollo, N. V., Maturana, M. I., Tong, W., Nayagam, D. A. X., Shivdasani, M. N. and Foroughi, J., *et al.* 2015. Soft, flexible freestanding neural stimulation and recording electrodes fabricated from reduced graphene oxide. *Advanced Functional Materials* 25: 3551–3559.

88. Banaee, H., Ahmed, M. and Loutfi, A. 2013. Data mining for wearable sensors in health monitoring systems: A review of recent trends and challenges. *Sensors* 13, 17472–17500.

89. Segev-Bar, M. and Haick, H. 2013. Flexible sensors based on nanoparticles. *ACS Nano* 7, no. 10: 8366–8378.

90. Clingan, H., Laidlaw, A., Tarakeshwar, P., Wimmer, M., García, A. and Mujica, V. 2018. Nanosensors for biomedical applications: A tutorial. *Semiconductor Nanotechnology*, 145–167.

91. Willner, I., Willner, B. and Katz, E. 2007. Biomolecule-nanoparticle hybrid systems for bioelectronic applications. *Bioelectrochem (Amsterdam, Netherlands)* 70, no. 1:2–11.

92. Rai, M., Gade, A., Gaikwad, S., Marcato, P. D. and Durán, N. 2012. Biomedical applications of nanobiosensors: The state-of-the-art. *Journal of the Brazilian Chemical Society* 23, no. 1: 14–24.

93. Shin, M. K. and Lee, B. 2012. Synergistic toughening of composite fibres by self-alignment of reduced graphene oxide and carbon nanotubes. *Nature Communications* 3, no. 1: 1–8.

94. Lee, J. H., Loya, P. E., Lou, J. and Thomas, E. 2014. Dynamic mechanical behavior of multilayer graphene via supersonic projectile penetration. *Science* 346: 1092–1096.

95. Bhowmik, S., Govindaraju, M., Ajeesh, G. and Sivakumar, V., Development of Light Weight Blast Proof Composite for Aviation, Space and Defence Structural Applications. U.S. Patent 2017410463972017.

96. Gao, Y., Cao, T., Cellini, F., Berger, C., Walter, A. H., Tosatti, E., Riedo, E. and Bongiorno, A. 2017. Ultrahard carbon film from epitaxial two-layer graphene. *Nature Nanotechnology* 13: 133–138.

97. Abu-Salah, K., Zourob, M., Mouffouk, F., Alrokayan, S., Alaamery, M. and Ansari, A. 2015. DNA-based nanobiosensors as an emerging platform for detection of disease. *Sensors* 15, no. 6: 14539–14568.

98. Malhotra, B. D. and Chaube, A. 2003. Biosensors for clinical diagnostics industry. *Sensors and Actuators B: Chemical* 91: 117–126.

99. Abu-Salah, K. M., Ansari, A. A. and Alrokayan, S. A. 2010. DNA-based applications in nanobiotechnology. *Journal of Biomedicine and Biotechnology* 2010: 1–15.

100. Peña-Bahamonde, J., Nguyen, H. N. and Fanourakis, S. K., *et al.* 2018. Recent advances in graphene-based biosensor technology with applications in life sciences. *Journal of Nanobiotechnology* 16, Article 75: 1–17.

10 Nanoscale Transistors

10.1 ISSUES WITH CMOS TECHNOLOGY SCALING

The invention of the field-effect transistor was filed as a patent for the first time in 1930 [1]. The Si-SiO_2-based metal oxide semiconductor (MOSFET) was introduced into practice after 30 years, in 1960 [2]. Due to high compatibility and lower cost, Si-based MOSFET has emerged as one of the promising devices in the electronic industry. MOSFET-based integrated circuits and systems have proved to be ultralow power alternatives. According to Moore's law, progress in the MOSFET field has followed an exponential growth in the last 25 years [3]. Since 1994, experts in the semiconductor industry have been reminded of the future technology requirements for fulfilling the requirements. In 1999, International Technology Roadmap for Semiconductors clearly predicted the issues with present MOSFET technology and provided future device requirements [4]. It had been predicted that by 2014, complementary metal oxide semiconductor (CMOS) technology would be out-of-date. As a result, the processor and RAM design are becoming challenging at lower channel lengths below 20 nm. Many researchers have highlighted the limitations of MOSFET and discussed many circuits and systems techniques to enhance performance [5–9]. Now the question is how the device limitations make energy-efficient system design possible? To conclude this we have highlighted some of the MOSFET limitations that are discussed below.

10.1.1 VELOCITY SATURATION AND MOBILITY DEGRADATION

With the MOSFET channel length scaling, the resultant electric field in the channel increases, which results in an increase in the velocity of charge carriers. Moreover, due to the high fields, the linear relationship between the electric field and the velocity of charge carriers will not exist. This is because when the electric field is very high, the electronic scattering rate increases resulting in an increase in the velocity of charge carriers. As a result, the overall transient time of charge carriers in the channel increases. It is identified that the MOSFET with lower channel lengths shows an increased average electron velocity than a bulk MOSFET. However, with scaled channel lengths [10], an increased electric field causes velocity saturation [Figure 10.1(a)] causing a decrease in the saturation current. Further, this effect causes the switching speed to drop in MOSFETs. In combination with the switching speed, the electron mobility of the device will also start degrading due to the scaled channel lengths, which is depicted in Figure 10.1(b).

10.1.2 TUNNELING LIMIT

The other main scaling limit of MOSFET is the tunneling current due to the weak gate insulator. The conventional insulator Si-SiO_2 will have high interfacing

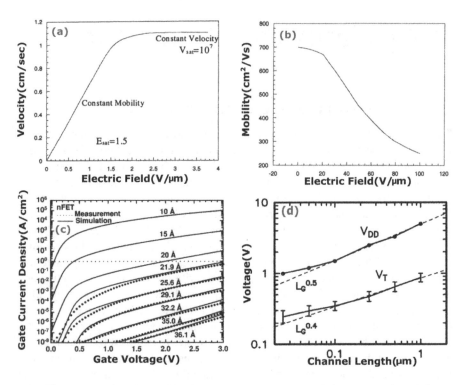

FIGURE 10.1 MOSFET behavior at scaled channel lengths. (a) Velocity saturation of charge carries with increased electric field. (b) Mobility of charge carriers with varying electric field. (c) Gate current density with increased Gate voltage. (d) Supply voltage and threshold reduction trends with varying channel lengths [10, 11].

capability, but with SiO_2, this leakage exceeds the requirements which are shown in Figure 10.1(c). This leakage effects are becoming the limitation to get lower power consumption in some applications like dynamic random access memory (DRAM) [11]. For higher channel length technologies, this will contribute a few milli watts to the overall chip dissipation. This leakage is especially problematic for ultralow power applications at lower channel lengths.

10.1.3 HIGH FIELD EFFECTS

Figure 10.1(d) depicts the scaling behavior of supply voltage and threshold voltage with the channel length [11]. From this, it can be observed that the supply voltage and threshold voltage have not scaled at the same rate as the length. As the power supply voltage has not been scaled in proportion to the channel length, the electric field strength has been increasing as CMOS devices have been scaled down. As a result, the oxide field has reached a maximum limit [12]. These fields can still reach higher values when the channel length goes into the nanometer regime. At such high fields, several undesirable effects occur, one such effect is the hot carrier effect.

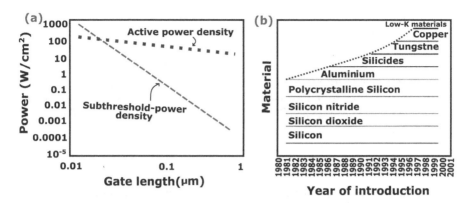

FIGURE 10.2 MOSFET (a) Power density trend with technology scaling (b) New materials with the year of introduction [13].

10.1.4 POWER LIMITATION

The power density of the MOSFET-based circuit is growing due to unexpected scaling limits of supply voltage [13]. In every integrated circuit (IC), the power consumption consists of mainly two components: dynamic power and static power consumption [14]. In the transition of device switching, there exists dynamic power consumption, while static power dissipation originates from the leakage source-drain current when the transistor is switched ON or OFF. Figure 10.2(a) shows both dynamic power density and static power density based on measured industrial data, as mentioned in Ref. [13], at a junction temperature of $T_j = 25°C$. The dynamic power density is found to be slightly more than 10 W/cm² at a gate length of 0.9 μm, while the static power density is a lower value. Moreover, both the power densities are becoming greater as gate length is reaching to smaller values. When gate length is becoming smaller that is at 20 nm gate length, the static power density is raised as high as dynamic power density.

10.1.5 MATERIAL LIMITATION

The emerging materials have been introduced to overcome the limitations of CMOS transistors. Figure 10.2(b) shows the introduction of different materials in the last three decades starting from 1980 to 2001 [13, 15]. These materials have been introduced to increase the manufacturability and reliability of devices. Material such as Silicon (Si), Silicon dioxide (SiO_2), Aluminum (Al), Copper (Cu) and Silicide are introduced to have enhanced dielectric constant (ε), carrier mobility (μ), carrier saturation velocity, breakdown field strength, and conductivity [16]. However, with the scaling of the transistor lengths, these devices reach their physical limits and show their unconventional behavior. For example, the SiO_2 reliability degrades as it becomes thinner and results in breakdown [17]. Even though Cu is less sensitive to electromigration than Al, the material is more susceptible to open defects when used as interconnect wires. As reported in [18], high-permittivity (k) materials have been used in 45 nm technology to replace silicon SiO_2 as the gate dielectric. Further, it is identified that the high-k materials can minimize the leakage current when the dielectric is made to become thinner to support physical scaling. However, with the new materials, there are other sets of problems with manufacturing tendency and temperature behavior that need to be studied before using them [19]. Several challenges are also reported in [20] such as appropriate tuning of metal work function, ensuring adequate channel mobility, gate stack integrity, and also the reliability of the material.

10.2   TUNNEL FET

This section presents an emerging tunnel field-effect transistor (TFET) device and explains its characteristics. Further, this section demonstrates the circuit design using TFET.

10.2.1   DEVICE STRUCTURE AND MODELS

Several researchers have proposed different types of TFET devices in the literature. Among several TFET proposals, III-V TFETs show promising behavior because of higher ON current [21–23]. This work explores LUT-based 20 nm InAs TFET Verilog-A model for circuit design and the physical structure of TFET is shown in Figure 10.3. Symbols for both N-channel TFET (NTFET) and P-channel TFET (PTFET) were created by deploying Verilog-A models using industry standard Cadence environment based on which TFET circuit designs are implemented. The model parameters and doping concentrations of the TFET device are detailed in Table 10.1 [22].

10.2.2   DEVICE CHARACTERISTICS

TFETs demonstrate peculiar electric characteristics, understanding of these characteristics is important to utilize TFETs for energy-efficient circuit design. TFETs can exhibit unique characteristics in four different quadrants by varying the gate-to-source voltage (V_{GS}) and drain-to-source voltage (V_{DS}) from negative to positive voltages. Figure 10.4 shows the different regions of N-channel TFET (NTFET) that possesses unique characteristics in four different quadrants.

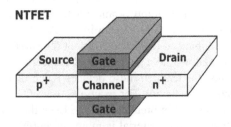

FIGURE 10.3   Physical *p-i-n* structure of *N*-channel TFET.

TABLE 10.1
TFET and FinFET Device Parameters [22]

Properties	NTFET	PTFET	Si FinFET
Gate length (L_g)	20 nm	20 nm	20 nm
EOT (HfO$_2$)	0.7 nm	0.5 nm	0.7 nm
Body thickness (T_b)	7 nm	0.7 nm	10 nm
Source doping concentration	4×10^{19} cm^{-3}	5×10^{18} cm^{-3}	1×10^{20} cm^{-3}
Drain doping concentration	2×10^{17} cm^{-3}	5×10^{19} cm^{-3}	1×10^{20} cm^{-3}
Gate work–function	4.85 eV	4.285 eV	4.55 eV
Hetero-junction band alignment	$E_{g,GaSb} = 0.845$ eV, $E_{g,InAs} = 0.49$ eV, $\Delta E_c = 0.439$ eV		–

10.2.2.1 TFET as ON Switch

Upon applying positive V_{GS} and V_{DS} voltages, the bands in the source and channel of TFET get aligned. Because of this, TFET with a reverse bias of *p-i-n* structure exhibits band-to-band tunneling from source to channel and the conduction is controlled by the gate voltage. In this region, TFET switches ON. Figure 10.5 shows the transfer characteristics of both NTFET and *N*-channel FinFET. At a supply voltage of 0.4 V, the TFET demonstrates 7 × higher ON current compared to the FinFET, whereas NTFET at beyond 0.4 V supply voltage shows lower current than FinFET.

10.2.2.2 Ambipolar Characteristics

When V_{GS} becomes negative, the tunneling takes place at the drain-channel junction that causes ambipolar current to flow from drain to source. Figure 10.6 presents the I_D–V_{GS} characteristics of *N*-channel TFET (NTFET) at $V_{DS} = 0.4$ V and

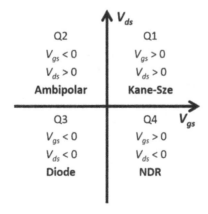

FIGURE 10.4   Operation of NTFET in different quadrants of V_{DS}-V_{GS}.

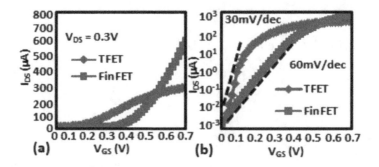

FIGURE 10.5 I_{DS}–V_{GS} characteristics of NTFET and Si FinFET (a) linear scale (b) log scale [24].

V_{GS} varying from −0.4 V to 0.4 V. From this, one can observe the presence of ambipolar current when the device is in the OFF state. At V_{GS} = −0.4 V, the NTFET shows an ambipolar current of 0.1 μA/μm. The NTFET and PTFET will be in the ambipolar region of operation for negative and positive values of V_{GS}, respectively.

10.2.2.3 Unidirectional Characteristics and *p-i-n* Forward Leakage

Figure 10.7 shows the I_D–V_{DS} characteristics of NTFET showing significant leakage current transport instead of the unidirectional current conduction with negative V_{DS}. When positive V_{DS} is applied, NTFET shows desired transistor behavior like conventional MOSFET. With negative V_{DS}, the TFET shows low leakage current with negative V_{DS}, because of the unidirectional property of the device that is due to the asymmetry of the device structure. When V_{DS} becomes more negative, there is an abrupt increase in leakage current. At V_{DS} = −0.4 V, NTFET exhibits a significant leakage current in the reverse direction (*i.e.* from source to drain), as shown in Figure 10.7.

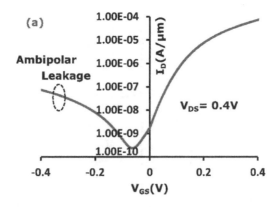

FIGURE 10.6 NTFET characteristics demonstrating ambipolar current.

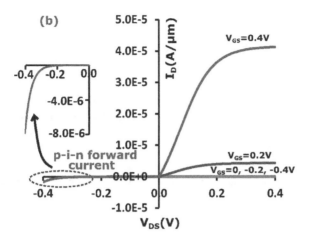

FIGURE 10.7 NTFET characteristics demonstrating *p-i-n* forward leakage current.

10.2.3 TFET-BASED CIRCUIT DESIGN

We have studied the basic building blocks using TFETs and benchmarked these designs with FinFETs. As the inverter is the basic building block of all digital circuits, we have introduced TFET inverter design that shows high energy efficiency compared to FinFETs.

10.2.3.1 TFET-Based Static Complementary Inverter Design

Figure 10.8 presents the TFET-based digital complementary inverter design comprising both PTFET and NTFET. The TFET-based inverter design is benchmarked against industry standard Si-FinFET-based inverter design. Figure 10.9 shows the transfer characteristics of both TFET and FinFET inverters. It can be observed that the TFET inverter characteristic is almost similar to an ideal inverter.

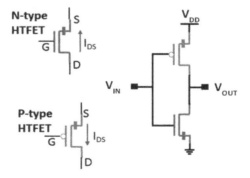

FIGURE 10.8 TFET static complementary inverter topology.

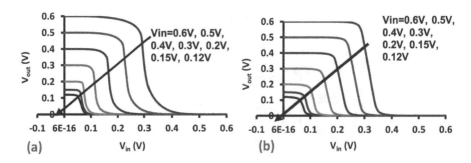

FIGURE 10.9   Inverter voltage transfer characteristics with supply voltage scaling (a) TFET and (b) FinFET [25].

10.2.3.2   TFET-Based Digital Buffer Design

Figure 10.10 demonstrates TFET and FinFET inverter-based 2-, 4-, and 6-stage buffers driving a load capacitance (C_L). TFET and FinFET buffers are designed considering identical parameters. Figure 10.11 shows the energy consumption of both TFET and FinFET buffers by varying load capacitances at a supply voltage of 0.2 V. It can be observed that TFET buffer designs achieve lower energy consumption than FinFET designs. It can be also observed that the FinFET buffers fail to drive the larger loads at a supply voltage of 0.2 V.

10.3   NEGATIVE CAPACITANCE FET

Negative capacitance FET (NCFET) as an emerging device candidate outperforms the CMOS technology and achieves lower power consumption in lower technology nodes [27, 28]. This section presents NCFET device structure and characteristics. Further, the NCFET-based logic circuits are benchmarked with CMOS technology.

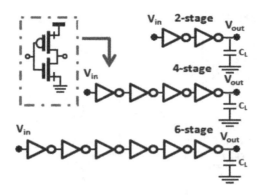

FIGURE 10.10   Complementary TFET inverter-based 2-, 4-, and 6-stage buffer designs [26].

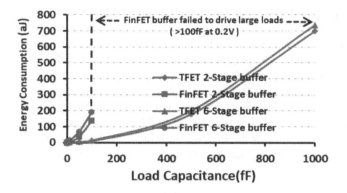

FIGURE 10.11 Energy consumption of 2-, 6-stage TFET, and FinFET buffer with varying load capacitance at 0.2 V supply voltage [26].

10.3.1 DEVICE STRUCTURE

A negative capacitance FET explores ferroelectric (FE) material as a gate oxide that exhibits subthreshold slope lesser than the fundamental Boltzmann limit of 60 mV/decade. In the recent past, numerous works have experimentally demonstrated the concept of negative capacitance. Moreover, NCFET with ferroelectric (FE) gate experimentally achieved lower subthreshold swing. The advancement in technology leads to scaling in the size of transistors and an increase in the density of chips that demonstrate the requirement of energy-efficient devices. These NCFET devices are more energy-efficient and also improves the on-current I_{on}. Figure 10.12 shows the structure of a typical NCFET device. NCFETs have recently come into spotlight as one of the steep switching devices that achieve the break-through in subthreshold slope (SS). The steep switching feature of an NCFET can be simply implemented in a conventional metal oxide semiconductor field-effect transistor (MOSFET) if the gate oxide layer of the MOSFET is replaced with a ferroelectric layer.

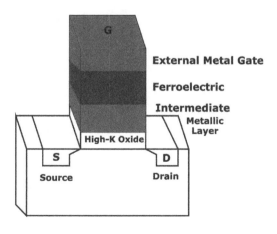

FIGURE 10.12 Structure of NCFET.

10.3.2   PRINCIPLE OF OPERATION

The subthreshold swing of the conventional MOSFET is limited due to the principle of thermionic injection. The 60 mV/decade subthreshold swing (*SS*) is the theoretical minimum value for any FET device, where the switching process involves the thermionic (temperature-dependent) injection of electrons over an energy barrier in order to flow current. This theoretical minimum is only achievable under the ideal conditions that cannot be satisfied. This sets a fundamental limit to the steepness of the transition slope from the "OFF" to the "ON" state. The subthreshold swing can be given by equation (10.1), where C_d and C_{ox} are the depletion and oxide capacitances, respectively.

$$SS = ln(10)\frac{KT}{q}\left[1+\frac{C_d}{C_{ox}}\right] \tag{10.1}$$

Various emerging devices are explored to lower *SS* below the thermionic limit of MOSFET by reducing the factor of C_d/C_{ox} in equation (10.1). Utilizing the concept of band-to-band tunneling, tunnel FETs have been proposed that exhibit lower subthreshold swing than MOSFET. Another alternative to reduce the *SS* is the introduction of negative capacitance on the gate of MOSFET.

10.3.3   LOW SUBTHRESHOLD SWING AND HIGH ON CURRENT

In the subthreshold region, the drain current behavior being controlled by the gate terminal is similar to the exponentially decreasing current of a forward-biased diode. Therefore, a plot of drain current versus gate voltage with drain, source, and bulk voltages fixed will exhibit approximately log–linear behavior in this MOSFET operating regime and its slope is the subthreshold region. The subthreshold slope for CMOS at 45 nm technology is 190 mV/decade and for NCFET device it is 58.4 mV/decade, as shown in Figure 10.13.

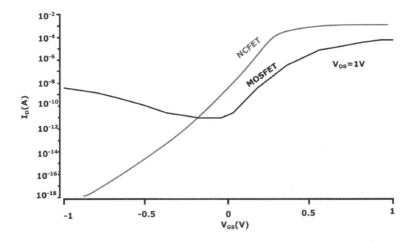

FIGURE 10.13   Subthreshold slopes for both NCFET and CMOS at 45 nm technology.

A typical experimental subthreshold swing for a scaled MOSFET at room temperature is ~70 mV/decade that is slightly degraded due to the short-channel MOSFET parasitic effects. A decade corresponds to a 10 times increase of the drain current I_D. A device characterized by steep subthreshold slope exhibits a faster transition between off (low current) and on (high current) states. These characteristics are divided into three ranges: (1) OFF range where $I_{DS} \leq I_{OFF}$, (2) ON region where the $I_{DS} \geq I_{ON}$, and (3) intermediate range where the transistor acts as an amplifier. For a given V_{DS}; the transfer characteristic (V_{gs} vs I_D) for the threshold level of gate voltage corresponding to the logarithmic scale of drain current (log I_D) is called I_{ON} and the least level of gate voltage when the minimum or only leakage conduction through the channel is available, the corresponding drain current logarithmic scale (log I_D) is called I_{OFF}. When comparing with CMOS at 45 nm technology with the NCFET device, the NCFET achieves the high On-current (I_{ON}).

10.3.4 HYSTERESIS CHARACTERISTICS

A negative capacitance FET is built by stacking a ferroelectric (FE) layer on the gate of a MOSFET, as shown in Figure 10.14. The figure shows the ferroelectric layer capacitance and the positive capacitance of the MOSFET as C_{FE} and C_{MOS}, respectively. A large C_{FE}/C_{MOS} ratio stabilizes the FE layer in the negative capacitance region, due to which the FE layer does not retain remnant polarization. As a result, the device exhibits voltage step-up action that can result in steep-switching behavior. This type of device is referred to as a negative capacitance field-effect transistor (NCFET), and is being explored by both academia and industry. As the FE layer thickness increases and the C_{FE}/C_{MOS} ratio decreases this results in a hysteretic behavior in an NCFET's transfer characteristic. The hysteretic behavior of NCFETs

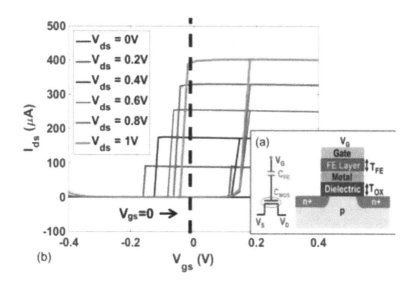

FIGURE 10.14   (a) Device structure (b) NCFET hysteresis characteristics [27].

TABLE 10.2

Propagation Delay and Energy Consumption of NCFET-Based Logic Gates at a Supply Voltage of 0.45 V

	Supply Voltage = 0.45 V; Load Capacitance = 500 fF							
Parameters	Inverter		NAND-Gate		NOR-Gate		Ex-OR-Gate	
	NCFET	CMOS	NCFET	CMOS	NCFET	CMOS	NCFET	CMOS
Delay (ns)	1.168	79.4	1.242	138.3	1.523	122.9	1.601	89.25
Energy (fJ)	0.2314	8.075	0.1402	6.56	0.2819	8.778	0.7608	6.518

is responsible for nonvolatility nature. This type of NCFET with hysteresis is called FeFET. Recent progress has been demonstrated with a promising performance of NCFET and utilized in embedded NV memories for low-cost internet-of-things (IoT) applications.

10.3.5 NCFET Device-Based Inverter and Digital Logic Design

The inverter using an NCFET device is more energy efficient compared to the CMOS device. Here, we have designed an NCFET inverter at 45 nm technology using the industry-standard Cadence Virtuoso tool. By using this device instead of CMOS, the delay and energy consumption have been reduced. Tables 10.2 and 10.3 summarize the propagation delay and energy consumption of NCFET-based logic gates at a supply voltage of 0.45 V and 0.6 V, respectively. It can be observed that the NCFET-based logic gates exhibit high speed and lower energy consumption compared to the CMOS logic gates.

10.4 CARBON NANOTUBE FET

This section presents carbon nanotubes (CNTs) and corresponding field-effect transistors. The characteristics of CNTFETs are discussed and the benefits of the device are highlighted.

TABLE 10.3

Propagation Delay and Energy Consumption of NCFET-Based Logic Gates at a Supply Voltage of 0.6 V

	Supply Voltage = 0.6 V; Load Capacitance = 500 fF							
Parameters	Inverter		NAND-Gate		NOR-Gate		Ex-OR-Gate	
	NCFET	CMOS	NCFET	CMOS	NCFET	CMOS	NCFET	CMOS
Delay (ns)	0.716	11.43	0.754	18.87	1.031	14.93	1.011	89.25
Energy (fJ)	0.416	2.067	0.207	1.714	0.518	1.896	1.449	3.85

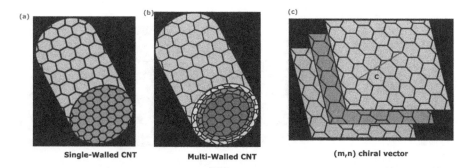

FIGURE 10.15   (a) A single-walled carbon nanotube, (b) A multi-walled carbon nanotube, and (c) Defining a chiral vector [29].

10.4.1 CARBON NANOTUBE

CNTs are obtained by rolling up graphene sheets with large aspect ratio as 10^3–10^5. Depending on the number of rolled-up graphene sheets, there are primarily two different types of CNTs:

- *Single-walled carbon nanotubes:* A single-walled carbon nanotube (SWCNT) is cylinder that can be obtained by rolling up a graphene sheet of hexagonal carbon rings.
- *Multi-walled carbon nanotubes:* A multi-walled carbon nanotubes (MWCNT) is a stack of graphene sheets as concentric circles.

The structures of both SWCNT and MWCNT are depicted in Figure 10.15. Each SWCNT is recognized by using a vector (n, m), called chiral vector. The chiral vector indicates the rolled-up direction of the graphene sheet and the diameter of the nanotube.

10.4.2 CARBON NANOTUBE FET

Single-walled carbon nanotubes exhibit high conductivity and excellent carrier mobility due to their small diameter. It has been experimentally demonstrated that these tubes can exhibit metallic or semiconducting characteristics depending on their chirality factor. Using semiconducting CNTs as a channel element, carbon nanotube FETs (CNTFETs) have been demonstrated [30]. A CNTFET refers to the field-effect transistor that utilizes a single CNT or an array of CNTs as the channel material instead of bulk silicon in the traditional MOSFET structures, as shown in Figure 10.16.

10.4.3 DEVICE CHARACTERISTICS

Figure 10.17 shows the I_D–V_{GS} and I_D–V_{DS} characteristics of CNTFET at the gate and drain voltages varying from 0 to 1 V, respectively. From the characteristics, it can be observed that the CNTFET exhibits the device characteristics that are similar to the MOSFET.

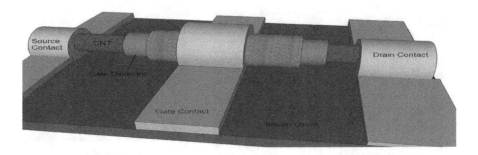

FIGURE 10.16   Structure of carbon nanotube FET.

It can also be observed that the threshold voltage of CNTFETs is varied according to the following relation:

$$D_{CNT} = \frac{\sqrt{3}a}{\pi}\sqrt{n^2 + m^2 + mn} \qquad (10.2)$$

$$V_{th} \approx \frac{E_g}{2e} = \frac{\sqrt{3}}{3}\frac{aV_\pi}{eD_{CNT}} \qquad (10.3)$$

where (n, m) is the chirality vector, $a = 2.46$ Å is the carbon to carbon atom distance, $V_\pi = 3.033$ eV is the carbon π–π bond energy in the tight bonding model, e is the unit electron charge, and D_{CNT} is the CNT diameter. Due to this, depending upon the chirality vector variation, the ON current and threshold voltage of CNTFET varies. This behavior can be clearly understood from the characteristics.

10.5   GRAPHENE NANORIBBON FET

Graphene has become a versatile material with unique properties that shows high performance when used as a channel material in a FET. This section introduces graphene, graphene-based nanoribbon, and graphene nanoribbon FETs (GNRFET).

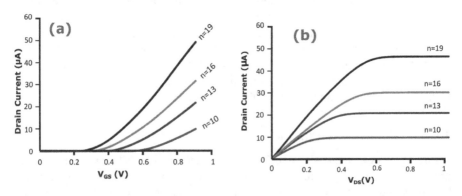

FIGURE 10.17   CNTFET (a) I_{DS}–V_{GS} characteristics and (b) I_{DS}–V_{DS} characteristics.

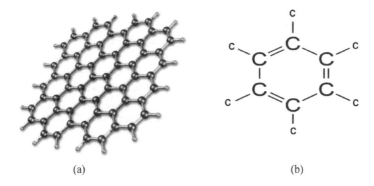

(a) (b)

FIGURE 10.18 Graphene crystal (a) lattice structure and (b) C–C bonding [31].

10.5.1 Graphene Structure and Properties

Graphene is a single layer of carbon atoms forming a "honeycomb lattice structure" that has attracted wide attention due to their unique characteristics. Each carbon atom is bonded to three other carbon atoms in the same plane. One of the bonds is a double bond and hence valency of the structure is satisfied, as shown in Figure 10.18.

The unique properties of graphene are as follows:

10.5.1.1 Mechanical Properties

The graphene is proved as robust material with the Young modulus of 1.0 TPa, the third-order elastic stiffness of −2.0 TPa, and the ultimate tensile strength of 130,000,000,000 Pascal. These strong mechanical properties of graphene make it superior to other commonly used materials such as silicon in electronics-based applications.

10.5.1.2 Electrical Properties

With the freely available electrons, graphene behaves as a semiconductor just like the silicon. However, since the cause of conduction in graphene is different from that of other semiconductor materials, it exhibits many unique properties. One such property is the electron mobility. The electron mobility of graphene in its pristine form is more than 200,000 cm^2/V s. The sheet resistance of graphene is about 30 ohms. Since graphene atoms are considered massless, they behave much similar to photons. The electrical conductivity of graphene is summarized in Table 10.4. It shows higher electrical conductivity compared to the other materials.

10.5.2 Graphene Nanoribbon FET

In order to understand the GNRFET, it is important to know about the graphene nanoribbon (GNR) properties. The detailed description of GNR and GNRFET is as follows:

10.5.2.1 Graphene Nanoribbon

Graphene nanoribbons (GNRs) are nanometer-sized structures that consist of single-layer graphene. Recently, GNRs have attracted wide attention and have been

TABLE 10.4

Summary of Electrical Conductivity of Different Materials [31]

Material	Electrical Conductivity (S m^{-1})
Graphene	$\sim 10^8$
Silver	63.0×10^6
Copper	59.6×10^6
Gold	45.2×10^6
Aluminum	37.8×10^6
Diamond	$\sim 10^{-11}$

demonstrated experimentally. As the structure of GNRs is similar to CNTs, they exhibit characteristics identical to CNTs. However, some theoretical studies demonstrate that the metallic or semiconducting feature in GNRs is different from that of CNTs. The electronic properties of nanoribbons depend on the dimensions of GNR such as the width and atomic geometry along the edge, namely zigzag GNR (ZGNR) and armchair GNR (AGNR). The structures of zigzag and armchair nanoribbons are shown in Figure 10.19.

An armchair GNR is metallic if the number of carbon atoms across its width N is $3p + 2$, where p is an integer. Zigzag GNR is always metallic independent of the number of carbon atoms.

10.5.2.2 Graphene Nanoribbon FET

Similar to the traditional silicon device, a GNRFET also has four terminals [32]. The undoped semiconducting GNR sheets are under the gate as the channel region, while heavily doped GNR segments are placed between the gate and the source/drain. As the gate potential increases, the device is electrostatically turned on or off via the gate. The current and voltage characteristics are the same as MOSFETs. The GNRFETs primarily provide a unique opportunity to control threshold voltage by the dimer lines or width. The structure of the GNRFET is shown in Figure 10.20.

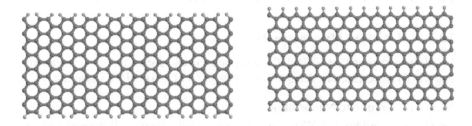

FIGURE 10.19 Armchair and zigzag structures of graphene.

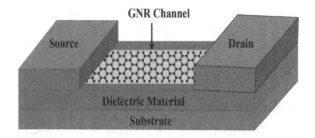

FIGURE 10.20 Graphene FET structure.

10.6 SPINTRONIC DEVICES

Exploring electronic spin instead of charge is popularly known as spintronics. Spintronic devices achieve nonvolatility, zero standby power, and higher density. Various novel materials and mechanisms have been developed to support the spintronics field. Due to the low standby power, the spin devices are used as nonvolatile memory applications. Apart from this, the spin devices are also utilized for various other applications for neuromorphic computing, hardware security, and logic design.

10.6.1 Principle of Operation

Giant magnetoresistance (GMR) is developed in a structure that contains a nonmagnetic material sandwiched between two ferromagnetic layers (shown in Figure 10.21) also known as spin value. It is proved that the magnetic layers with magnetization antiparallel to each other exhibit higher resistance than the parallel magnetization state. The GMR is defined as

$$GMR = \frac{R_{AP} - R_P}{R_P} \tag{10.4}$$

where R_P is the resistance in parallel magnetization state which is low in value.

R_{AP} is the resistance in antiparallel magnetization state which is high in value. As the GMR phenomenon is very weak, a tunnel barrier is inserted between free layers, which results in a strong change in resistance called tunnel magnetoresistance (TMR).

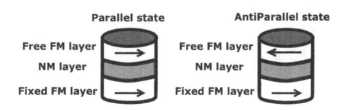

FIGURE 10.21 Structure of MTJ.

This TMR is observed in new devices called a magnetic tunnel junction (MTJ). The TMR effect introduces the novel class of nonvolatile memory technology named as magnetoresistive random access memories (MRAMs) that store the data in the form of the resistance state of MTJ.

10.6.2 Spin-Based Devices

There are different spintronics devices [33] available in the literature, those are:

- STT MTJs
- GSHE-based devices
- All spin logic (ASL) devices

10.6.2.1 STT MTJs

Magnetic tunnel junctions (MTJs) are multilayer stacked structures that explore spin-dependent quantum-mechanical tunneling. The MTJ consists of three layers: ferromagnetic free layers, tunnel barrier, and ferromagnetic fixed layer, as shown in Figure 10.21. Generally, the tunnel barrier used in MTJs is an oxide material such as MgO_X, TiO_X, and AlO_X. In this MTJ device, the switching between two states is achieved using MJT current, as shown in Figure 10.22.

When a sufficiently large current flows from layers FM1 to FM2, the resistance of MTJ switches from a higher value to a lower value. In this condition, both the layers are parallel to each other and the current used to make this is called parallelizing current. Similarly, when a large current flows from the layers FM2 to FM1, the resistance of MTJ switches from a lower value to a higher value. In this condition, both the layers are antiparallel to each other.

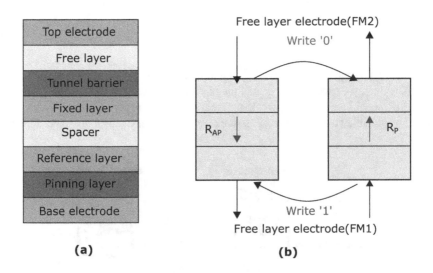

(a)   (b)

FIGURE 10.22 (a) Architecture and (b) operation of STT MTJ [33].

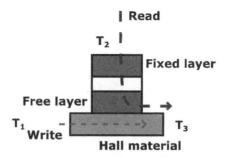

FIGURE 10.23   Structure of GSHE-based MTJ.

10.6.2.2   GSHE-Based Devices

Figure 10.23 shows the architecture of the giant-spin Hall-effect (GSHE)-based MTJ devices wherein a nonmagnetic layer carries a charge current. The MTJ structure is placed on the top of a nonmagnet layer, as shown in Figure 10.23. The magnetization of the ferromagnetic-free layer is varied by using the direction of current flow. The GSHE has become a novel three-terminal device where these devices have separate read and write current paths.

10.6.2.3   ASL Devices

The ASL devices (ASLD) can perform two operations: storing the information and processing the information. The primary advantage of this device compared to the CMOS technology is self-logic capability. The ASLD consists of nanomagnets (NMs) that can be switched into different stable states using a spin-polarized current. The nanomagnets can be built over an NM channel that carries the spin-polarized current. The structure and operation of an ASLD are shown in Figure 10.24. A charge current flows from the injector FM to ground contact through a channel to produce a nonequilibrium concentration of spin-polarized electrons beneath the injector. This nonequilibrium concentration of spin-polarized electrons diffuses through the channel to reach the detector FM with some losses due to the spin relaxation. The diffused electrons generate spin-polarized current and exert STT on the detector to eventually determine its final state. The magnitude of spin-polarized current through the channel depends on the potential applied to the injector FM.

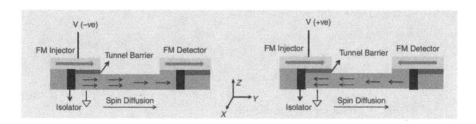

FIGURE 10.24   Structure of ASL device.

10.7   SUMMARY

This chapter can be summarized as follows:

- This chapter discussed the emerging device technologies that can out-perform the performance of the CMOS technology at lower technology nodes.
- First, we discussed the issues with present CMOS technology and their limitations.
- We introduced the emerging semiconductor devices and properties. Various emerging devices, namely tunnel FET, negative capacitance FET, carbon nanotube FET, graphene nanoribbon FET, and spintronics devices.
- Next, different characteristics of emerging devices were discussed, which are used for energy-efficient circuit design.
- Finally, the emerging device-based circuits designs were discussed, and benchmarking was performed against CMOS technology.

10.8   MULTIPLE-CHOICE QUESTIONS

1. Principle of operation of tunnel FET
   a. Band-to-band tunneling
   b. Drift
   c. Diffusion
   d. Thermionic emission

2. Tunnel FET exhibits
   a. Lower subthreshold swing
   b. Higher delay
   c. Higher subthreshold swing
   d. None

3. Hysteresis characteristics can be seen in
   a. NCFET
   b. TFET
   c. CNTFET
   d. MOSFET

4. p-i-n leakage is a characteristic of
   a. NCFET
   b. TFET
   c. CNTFET
   d. MOSFET

5. Leakage of MOSFET increases with
   a. Reduction in channel length
   b. Increase in channel length

   c. Independent of channel length
   d. None

6. Advantage of spintronic devices
   a. Low static power consumption
   b. High static power consumption
   c. High speed
   d. None

7. Zigzag GNR acts as
   a. Metal
   b. Semiconductor
   c. Insulator
   d. None

8. Ambipolar current can be seen in
   a. TFET
   b. CNTFET
   c. Both a and b
   d. None

9. GSHE is
   a. Giant-spin Hall-effect
   b. Giant-spin hysteresis-effect
   c. Giant-semiconductor hysteresis-effect
   d. Giant-semiconductor Hall-effect

10. FinFET uses which materials
   a. Silicon
   b. InAs
   c. Graphene
   d. None

10.9 SHORT ANSWER QUESTIONS

1. What is band-to-band tunneling?
2. What is the principle of operation of NCFET?
3. What is the principle of operation of MTJ?
4. What are the types of graphene structures?
5. What is the subthreshold swing limit of MOSFET?
6. What is the condition for hysteresis in NCFETs?
7. What are the advantages of ASL devices?
8. What is the difference between CNT and graphene nanoribbons?
9. What is *p-i-n* forward leakage of TFET?
10. What are the advantages of spintronics-based memory?

10.10   LONG ANSWER QUESTIONS

1. What are the different types of spintronic devices? Explain.
2. How band-to-band tunneling is different from thermionic emission? What are the benefits of band-to-band tunneling when used in TFET?
3. What is the concept of negative capacitance? How it will be used to improve the performance of FET?
4. Explain the different characteristics of tunnel FET.
5. Explain the principle of operation of carbon nanotube and graphene FET.

REFERENCES

1. Lilienfeld, J. E. 1930. Method and apparatus for controlling electric currents, U.S. Patent 1 745 175.
2. Kahng, D. and Atalla, M. M. 1960. Silicon–silicon dioxide field induced surface devices, presented at the IRE Solid-State Device Res. Conf., Pittsburgh, PA.
3. Bondy, P. K. 1998. Moore's law governs the silicon revolution. *Proceedings of the IEEE* 86: 78–81.
4. Semiconductor Industry Association (SIA). 1999. *International Technology Roadmap for Semiconductors* (1999 edition), SIA, San Jose, CA.
5. Davari, B., Dennard, R. H. and Shahidi, G. G. 1995. CMOS scaling for high-performance and low-power—the next ten years. *Proceedings of the IEEE* 89: 595–606.
6. Taur, Y., Buchanan, D., Chen, W., Frank, D., Ismail, K., Lo, S. H., Sai-Halasz, G., Viswanathan, R., Wann, H. J. C., Wind, S. and Wong, H. S. 1997. CMOS scaling into the nanometer regime. *Proceedings of the IEEE* 85: 486–504.
7. Asai, S. and Wada, Y. 1997. Technology challenges for integration near and below 0.1 m. *Proceedings of the IEEE* 85: 505–520.
8. Sugii, T., Momiyama, Y., Deura, M. and Goto, K. 1999. MOS scaling beyond 0.1 m. In *Silicon Nanoelectronics Workshop*, pp. 60–61.
9. Wong, H. S. P., Frank, D. J., Solomon, P. M., Wann, H. J. and Welser, J. 1999. Nanoscale CMOS. *Proceedings of the IEEE* 87: 537–570.
10. Mohab, A., Allam, M. and Elmasry, M. 2002. Impact of technology scaling on CMOS logic styles. *IEEE Transactions on Circuits and Systems II: Analog and Digital Signal Processing* 49, no. 8: 577–588.
11. Frank, D. J., et al. 2001. Device scaling limits of Si MOSFETs and their application dependencies. *Proceedings of the IEEE* 89, no. 3: 259–288.
12. Taur, Yuan, et al. 1997. CMOS scaling into the nanometer regime. *Proceedings of the IEEE* 85, no. 4: 486–504.
13. Nowak, E. J. 2002. Maintaining the benefit of CMOS scaling when scaling bogs down. *IBM Journal of R&D* 46, no. 2: 169–180.
14. Frank, D. J. 2002. Power constraint CMOS scaling limits. *IBM Journal of R&D* 46, no. 2: 235–244.
15. Waser, R. 2005. *Nanoelectronics and Information Technology: Advanced Electronics Materials and Novel Devices* (2nd edition), Wiley-VCH, New York, NY.
16. Wang, Y., et al. 1998. The challenges for physical limitation in Si microelectronics. *Proceedings of the 5th International Conference on Solid-State and Integrated Circuit Technology*, 1998, pp. 25–30.
17. Iwai, H. 2004. CMOS Scaling for sub-90 nm to sub-10 nm. *Proceedings of the 17th International Conference on VLSI Design (VLSID04)*, pp. 30–35.

18. Tyagi, S., 2007. Moore's law: A CMOS scaling perspective. In *Proceedings of 14th International Symposium on the Physical and Failure Analysis of Integrated Circuits, 2007 (IPFA 2007)*, pp. 10–15.

19. Buchanan, D. A. 1999. Scaling the gate dielectric: Materials, integration and reliability. *IBM Journal of R&D* 43, no. 3: 245–264.

20. International Technology Roadmap for Semiconductor. 2007. Edition Executive Summary. http://www.itrs.netiLinks/2007ITRS/ExecSum2007.pdf

21. Hao, L., Trond, Y. and Alan, S. 2015. Universal TFET model implementation in Verilog-A, version 1.6.8 (https://nanohub.org/publications/31).

22. Liu, H., Narayanan, V. and Datta, S. 2015. III-V Tunnel FET model, version 1.0.1 (https://nanohub.org/publications/12/2).

23. Sant, S., Moselund, K., Cutaia, D., Schmid, H., Borg, M., Riel, H. and Schenk, A. 2016. Lateral InAs/Si p-type tunnel FETs integrated on Si—Part 2: Simulation study of the impact of interface traps. *IEEE Transactions on Electron Devices* 63, no. 11: 4240–4247.

24. Aditya, J., Sadulla, S. and Vaddi, R. 2016. Exploiting the steep subthreshold slope characteristics of tunnel transistors for wide tuning range voltage controlled ring oscillator (VCRO) design at scaled supply voltages down to 150 mV. In *2016 3rd International Conference on Emerging Electronics (ICEE)*, pp. 1–2.

25. Vallabhaneni, H., Japa, A., Shaik, S., Krishna, K. S. R. and Vaddi, R. 2014. Designing energy efficient logic gates with Hetero junction Tunnel fets at 20 nm. In *2014 2nd International Conference on Devices, Circuits and Systems (ICDCS)*, pp. 1–5.

26. Aditya, J., Harshita, V. and Vaddi, R. 2017. Exploiting characteristics of steep slope tunnel transistors towards energy efficient and reliable buffer designs for IoT SoCs. *International Symposium on VLSI Design and Test,* Springer, Singapore, pp. 259–269.

27. Yin, X., Chen, X., Niemier, M. and Hu, X. S. 2018. Ferroelectric FETs-based non-volatile logic-in-memory circuits. *IEEE Transactions on Very Large Scale Integration (VLSI) Systems* 27, no. 1: 159–172.

28. Tian, H., Li, Y. X., Li, L., Wang, X., Liang, R., Yang, Y. and Ren, T. L. 2019. Negative capacitance Black phosphorus transistors with low SS. *IEEE Transactions on Electron Devices* 66, no. 3: 1579–1583.

29. Liu, X., Lee, C., Zhou, C. and Han, J. 2001. Carbon nanotube field-effect inverters. *Applied Physics Letters* 79, no. 20: 3329–3331.

30. Chen, Z., Farmer, D., Xu, S., Gordon, R., Avouris, P. and Appenzeller, J. 2008. Externally assembled gate-all-around carbon nanotube field-effect transistor. *IEEE Electron Device Letters* 29, no. 2: 183–185.

31. Fan, Y., Wang, L., Li, J., Li, J., Sun, S., Chen, F., Chen, L. and Jiang, W. 2010. Preparation and electrical properties of graphene nanosheet/Al_2O_3 composites. *Carbon* 48, no. 6: 1743–1749.

32. Moon, J. S., Seo, H. C., Stratan, F., Antcliffe, M., Schmitz, A., Ross, R. S., Kiselev, A. A., Wheeler, V. D., Nyakiti, L. O., Gaskill, D. K. and Lee, K. M. 2013. Lateral graphene heterostructure field-effect transistor. *IEEE Electron Device Letters* 34, no. 9: 1190–1192.

33. Verma, S., Kulkarni, A. A. and Kaushik, B. K. 2016. Spintronics-based devices to circuits: Perspectives and challenges. *IEEE Nanotechnology Magazine* 10, no. 4: 13–28.

MCQ Answers

CHAPTER 1 SEMICONDUCTOR PHYSICS AND DEVICES

Sr. No.	Ans	Sr. No.	Ans
1.	d	6.	b
2.	a	7.	d
3.	a	8.	a
4.	c	9.	a
5.	b	10.	b

CHAPTER 2 VLSI SCALING AND FABRICATION

Sr. No.	Ans	Sr. No.	Ans	Sr. No.	Ans	Sr. No.	Ans
1.	c	6.	b	11.	b	16.	a
2.	d	7.	a	12.	c	17.	c
3.	b	8.	b	13.	c	18.	d
4.	a	9.	b	14.	a	19.	b
5.	b	10.	d	15.	b	20.	a

CHAPTER 3 MOSFET MODELING

Sr. No.	Ans	Sr. No.	Ans	Sr. No.	Ans	Sr. No.	Ans
1.	b	4.	b	7.	d	10.	a
2.	c	5.	c	8.	c		
3.	b	6.	a	9.	b		

CHAPTER 4 COMBINATIONAL AND SEQUENTIAL DESIGN IN CMOS

Sr. No.	Ans	Sr. No.	Ans
1.	d	6.	a
2.	b	7.	c
3.	a	8.	c
4.	a	9.	a
5.	d	10.	b

CHAPTER 5 ANALOG CIRCUIT DESIGN

Sr. No.	Ans	Sr. No.	Ans
1.	a	6.	a
2.	d	7.	d
3.	a	8.	a
4.	a	9.	b
5.	b	10.	a

CHAPTER 6 DIGITAL DESIGN THROUGH VERILOG HDL

Sr. No.	Ans	Sr. No.	Ans
1.	a	6.	c
2.	d	7.	a
3.	c	8.	a
4.	a	9.	a
5.	a	10.	a

CHAPTER 7 VLSI INTERCONNECT AND IMPLEMENTATION

Sr. No.	Ans	Sr. No.	Ans	Sr. No.	Ans	Sr. No.	Ans
1.	b	5.	d	9.	d	13.	c
2.	a	6.	a	10.	a	14.	d
3.	b	7.	d	11.	c	15.	c
4.	b	8.	d	12.	a		

CHAPTER 8 VLSI DESIGN AND TESTABILITY

Sr. No.	Ans	Sr. No.	Ans	Sr. No.	Ans	Sr. No.	Ans
1.	c	5.	b	9.	d	13.	c
2.	b	6.	d	10.	a	14.	c
3.	d	7.	a	11.	a	15.	b
4.	d	8.	c	12.	b	16.	d

CHAPTER 9 NANOMATERIALS AND APPLICATIONS

Sr. No.	Ans	Sr. No.	Ans	Sr. No.	Ans	Sr. No.	Ans
1.	b	5.	b	9.	a	13.	c
2.	b	6.	b	10.	c	14.	d
3.	d	7.	c	11.	d	15.	d
4.	d	8.	b	12.	b		

CHAPTER 10 NANOSCALE TRANSISTORS

Sr. No.	Ans	Sr. No.	Ans
1.	a	6.	a
2.	a	7.	a
3.	a	8.	c
4.	b	9.	a
5.	a	10.	a

Index

323